Cosmic Catastrophes

Cosmic Catastrophes

Clark R. Chapman
and
David Morrison

Plenum Press • New York and London

Library of Congress Cataloging in Publication Data

Chapman, Clark R.
Cosmic catastrophes / Clark R. Chapman and David Morrison.
p. cm.
Includes bibliographical references and index.
ISBN 0-306-43163-7
1. Astronomy. 2. Disasters. I. Morrison, David 1940- . II. Title.
QB43.2.C47 1989 88-26819
523—dc19 CIP

First Printing—April 1989
Second Printing—February 1990

Plenum Press is a Division of Plenum Publishing Corporation
233 Spring Street, New York, N.Y. 10013

Printed in the United States of America

Preface

We are privileged to live in an extraordinary time. Space-age methods of science are revolutionizing human understanding of the nature of the cosmos and our place in it. One of the most profound developments in science during the past two decades has been the recognition that sudden, immense, often unpredictable forces—in short, catastrophes—have helped to shape the stars, the planets, and the Earth. Although catastrophes are extremely rare, their effects are so enormous that many of the traits of our world, and of the solar system, may be due to them. This new perspective differs from the traditional uniformitarian view that the world is shaped almost entirely by the familiar, continually acting processes that we can observe every day. Although many catastrophic events and processes were more common billions of years ago than today, the threat of cosmic-scale catastrophes remains very immediate. Indeed, civilization as we know it could be changed or even doomed within the lifetimes of our grandchildren by collision of the Earth with a comet, by nuclear war, or by human-wrought climate changes.

Until recently, talk of catastrophes in the history of the Earth has been synonymous with pseudoscience, that wild, nonscientific blend of superstition and hype that fascinates so many people. Recently, however, scientists themselves have begun to understand the significance of catastrophes in the formation of the landforms of the planets and in the evolution of life on Earth. In this book, we try to describe the evidence for this new view of the cosmos. We approach our topic nontechnically, but with enough philosophical rigor that other scientists, too, may find our discourse interesting.

Some of our topics—supernova explosions, colliding worlds, the mass extinctions at the end of the Cretaceous, and nuclear winter—are catastrophes by any definition. But we have also broadened our perspective to include other cosmic or planetary phenomena that involve rapid or profound changes in planetary environments, even though not of a specifically catastrophic nature, such as ice ages, the global greenhouse effect, and the ultimate death of the Sun.

We have profited from discussions with many of our colleagues and friends, quite a few of whom have read one or more chapters in draft stage. We wish especially to thank Janet Morrison and Lonny Baker for their many helpful comments about much of the manuscript. We have also benefitted from the numerous suggestions of our editors at Plenum, Linda Greenspan Regan and Victoria Cherney, and we thank them also for their patience during the drawn-out final stages of this project.

Clark R. Chapman and David Morrison

Tucson and Honolulu

Contents

CHAPTER 1

Prologue: Danger from the Skies

The scene is Siberia, in the great taiga forests below the Arctic Circle; the date is June 30, 1992. Here, facing a hostile United States, the Soviet Union has constructed one of its most sophisticated defense bases. Huge radars scan north and east seeking intruders into Soviet airspace, including possible missiles launched from United States bases or from submarines beneath the arctic ice. The latest MIG all-weather fighters are poised in blast-resistant shelters along a series of military runways. Surface-to-air (SAM) missiles cluster around this giant complex. Although no immediate international crisis is evident, the thousands of Soviet military and civilian workers at this space-age facility are alert, for they know that they would be among the first targets in a nuclear war. The very survival of their nation could depend on their quick response to a challenge from the American imperialists.

With their powerful phased-array systems, the Soviet radar operators are able to scan much of near-Earth space. They routinely track thousands of individual satellites and fragments of space junk. Yet there is no warning at 7:30 on this sunny morning when suddenly a column of fire descends from the east. People standing outside gaze up in fright and awe at a fireball, as bright as the Sun, speeding silently toward them. Before most of them can react or dive for shelter, the fireball explodes in a 10-megaton airburst about 20,000 feet above the military base. The first evidence of the blast for those indoors comes as power surges incapacitate the radars and computers. Communications with Moscow and the Soviet High Command are cut instantly. A few seconds later the shock wave from the explosion strikes, smashing buildings, toppling the 200-foot-high radar antennas, and scattering airplanes about like crumpled toys. Black-and-red towers of smoke and flame pour from ruptured fuel storage tanks. The Siberian forest itself is flattened for thousands of square miles, the trees stripped of branches and leaves and left scattered like matchsticks all pointing away from the blast. Of the personnel who lived at the base, only a handful survive among the raging fires that devour the shattered ruins. As orbiting satellites reveal the destruction of the base, desperate Soviet military commanders prepare to retaliate against the Americans before attacks can be launched against other vital defense positions.

What was the source of this devastating explosion? Has the United States developed a secret weapon capable of penetrating Soviet airspace undetected? Was there an accident, or has some military madman triggered global nuclear war? Will the United States itself launch an all-out attack before the Soviet missiles leave their silos? Decisions must be made in a matter of seconds, yet no one really knows what has happened.

This scenario sounds like science fiction or the beginning of some apocalyptic novel of nuclear war. Unfortunately, the explanation is both more prosaic and more frightening, for this explosion actually took place over Siberia just as we have described it—except that the date was June 30, 1908, and not 1992. At that time, the desolate stretch of Siberia near the Tunguska River was empty wilderness, and there were no fatalities since no one was present at the explosion site. There was only one literate witness, a trader at a post about 70 miles away, and the explosion even at that distance was sufficient to knock him off his chair. But the blast wave was large enough to be detected by instruments all around the world, and the shattered forest had still not recovered when Soviet scientific expeditions surveyed the area 20 years later. The only invention in our story is the Soviet defense base—but there are such bases in Siberia today, and it is sobering to imagine the world's reaction if such a 10-megaton blast took place there today, or anywhere in the world.

The Tunguska explosion of 1908 was caused by the collision of a fragment of cosmic debris with the Earth's atmosphere; it was probably a mass of ice and rock about 300 feet in diameter. The space between the planets is filled with such material, much of it in the form of objects a lot smaller than the Tunguska projectile. Every day small fragments of rock and metal plunge into the Earth's atmosphere and survive to land as meteorites. These "stones from heaven" are typ-

Devastated Siberian forest from the 1908 Tungaska impact, the most violent impact catastrophe of the 20th century. (Novosti)

ically only a few inches across, and so far there is no record of a human ever being killed by one. However, we also know that objects as large as the one that exploded over Tunguska are found throughout interplanetary space, remnants of the formation of the solar system.

Large as it was, the Tunguska projectile was too small to have been detected telescopically before it struck the Earth. We cannot be sure how many such objects there are, lurking between the planets. But astronomers have direct evidence of still larger potential projectiles, since these are occasionally revealed to our telescopes. These so-called "Earth-approaching asteroids" are as much as 10 miles in diameter, and they are capable of inflicting vastly greater damage than occurred at Tunguska. Comets, which are composed of both rock and ice and which can appear from the depths of space at any time, are also capable of striking the Earth. The consequences of such events in the past history of our planet are only now becoming apparent, and we have hardly begun to think of the damage they could inflict in the future.

The most dramatic effects of the impact of an Earth-approaching asteroid are associated with a major break in the fossil record of terrestrial life that occurred 65 million years ago. At that time, at the boundary between rocks of the older Cretaceous and newer Tertiary periods of geological history, more than half of the total species on our planet became extinct. There is now clear evidence that this discontinuity was caused by the collision of an asteroid or comet with the Earth. The pulverized remains of this celestial interloper, together with millions of tons of terrestrial rock dust resulting from the impact, can be found today in a layer of clay deposited all over the globe at the boundary between the rocks of the Cretaceous and Tertiary periods. This one event, at least, was of such a magnitude as to alter fundamentally the evolution of life on our planet. One naturally wonders how often such events have taken place and in what other ways they have helped shape the history of the Earth.

Actually, the role of cosmic impacts in the history of the Earth should have been obvious long ago—or at least the evidence of such events seems clear enough in hindsight. When Galileo turned his first telescope toward the Moon in 1610 and observed it 30 times magnified, he saw that its surface was covered with shallow circular depressions. Later astronomers called these features "craters," the Greek word for "cup." Astronomers estimate that 30,000 craters can be seen

The heavily cratered highlands of the Moon record the period of heavy bombardment that marked the first 500 million years of lunar history. (NASA/Johnson Space Center)

on the Moon with a large telescope, and uncounted others could be added from spacecraft surveys. Where did all of these craters come from? The answer today seems obvious: They are scars produced by the impacts of cosmic debris. Yet as little as a generation ago many geologists and astronomers thought that the craters of the Moon were of volcanic origin.

The Moon is our nearest neighbor in space, and it and the Earth are subject to the same rain of interplanetary debris. While it is true that the Earth has an atmosphere and the Moon does not, our atmosphere provides little protection against large impacts of the sort that concern us here. Thus, understanding the origin of the lunar craters is closely connected to an appreciation of the role of impacts on our own planet. Because scientists failed until recently to recognize the impact origin of lunar craters, they also misjudged the geological history of the Earth. For the most part, it never occurred to them to look for the terrestrial scars—much eroded—of a similar history of cosmic impacts.

The very term "catastrophe" carries a negative connotation to many scientists. It suggests the irrational or unpredictable and conjures up images of the "acts of God" much beloved by newspaper writers. Science, after all, deals exclusively with a universe of natural laws and causality. Since the validity of its theories depends on reproducibility—in the laboratory or in nature itself—science is not well equipped to interpret random, chaotic, or unique events. Geologists, in particular, have developed a distaste for explanations of natural events that seem to require arbitrary and isolated causes. Such a reaction is understandable, given the history of this science, which struggled long to free itself from the tradition of a recent creation and a worldwide flood, based on a literal interpretation of the first five chapters of Genesis. One of the triumphs of 19th-century science was the recognition of the great age of the Earth and the fact that terrestrial geology (much of it, anyway) could be understood without invoking catastrophes. Having once established that the visible landscape of our planet could be formed over vast epochs of time by the slow but persistent application of the currently observed forces of uplift, erosion, and sedimentation, these scientists were reluctant to consider the role of cosmic impacts or other catastrophes. Only in the past decade has evidence grown to the point where the existence—and importance—of such catastrophes can no longer be denied.

As we have come to appreciate the role of catastrophes in the history of the Earth, we have also become sensitized to the more subtle evidence of major breaks in the history of other worlds. If the Earth and Moon have been bombarded by thousands of fragments of cosmic debris, what about the other members of the solar system? As spacecraft have extended our vision outward, they have radioed back evidence that immense impacts and other disruptions have punctuated the history of these worlds as well. Indeed, in many ways the exploration of other planets has been a humbling experience, for we have seen that the Earth is overshadowed in many ways. Other planets bear larger impact scars than does our own. The volcanoes of Mars are far larger than those of Earth, and even a moon of Jupiter—Io—experiences a level of volcanic activity many times greater than that of our own planet. A few moons and planets seem to have experienced impacts of such incredible violence that they were disrupted and dispersed, or they were nearly broken apart and fell back together in a jumbled mass of fragments. Elsewhere in the astronomical universe, also, we see evidence of violence—colliding galaxies, exploding stars, and new generations of worlds forming from the shattered remnants of their predecessors.

Scientists now recognize that the universe is dynamic and tumultuous—a place of change, often violent and catastrophic. Almost unconsciously, there has occurred a revolution in our thinking about nature. It is not the universe that has changed during the past generation, but our perception of it. Whether our current ideas will still be in vogue a generation from now, we cannot know, but it seems clear that we will never return to the concept of a nearly static universe, proceeding on a slow and majestic evolution unbroken by the rare, the unexpected, and the violent events that are the subject of this book.

CHAPTER 2

Impacts on the Earth

Only within recent years have we realized the threat to the Earth posed by the impacts of comets and asteroids. In 1908, at the time of the Tunguska event, no asteroids had been discovered in orbits that brought them close to the Earth. Of all the objects in space, only comets seemed capable of colliding with the Earth, but there was little appreciation of the danger they represented. At that time comets were commonly thought to be made of insubstantial stuff, rather like a loose pile of sand or gravel. The idea that comets might have solid, rocky nuclei was not to emerge until 1950, and only in 1986 did the first spacecraft come close enough to a comet (Halley's Comet) to photograph its solid nucleus.

At the time of the Tunguska event, and for many years afterward, this spectacular explosion—one of the greatest in human history—was relegated to the dubious category of curious but unverified phenomena. The first scientific expedition to investigate the area did not take place until 1927, when L. Kulik, curator of the meteorite collection of the Mineralogical Museum in Leningrad, led a small group on foot over frozen reindeer tracks to the site of the explosion. They reached an area of devastated forest, but they were forced back by the weather. Kulik returned later that year, and again in 1929, but no crater or other direct evidence of the impacting object was found. Had it not been for the blast wave, recorded on sensitive atmospheric instruments that had been installed at laboratories around the world only a few years before 1908, we would probably not even know of the Tunguska event today.

One reason that scientists took so long to recognize the possibilities of impacts on the Earth was our ignorance of the existence of projectiles with orbits that crossed our own. Even more important, however, was the obvious lack of direct evidence. With the exception of the Tunguska event, no one had witnessed a major impact. Nor were the scars of impacts to be recognized on the surface of the Earth (a consequence, scientists now realize, of our planet's active geological forces, which erode impact craters as rapidly as they form and soon erase them entirely). Such craters as are apparent on our planet are nearly all of volcanic origin, and not the result of impacts. Geologists were naturally concerned with the origin of the primary structures on our planet, such as its continental masses and folded mountain ranges, which are complicated enough to understand. Few had any desire to speculate about the effects of large impacts when there was little or no evidence of them in the landscape.

Although impact structures are rare on the Earth, it is possible to find a few of them, if we know what to seek. In particular, there is one undeniable impact crater on the Earth, and that is Meteor Crater in Arizona. This structure, which looks very much like the ubiquitous lunar craters, has become the best studied impact feature on Earth.

* * *

Many tourists driving across northern Arizona on Interstate 40 have seen the turnoff to Meteor Crater. Along this desolate stretch of road, between Petrified Forest National Park and the Painted Desert to the east and the volcanic San Francisco peaks to the northwest near Flagstaff, there is little else to catch one's eye. Yet most travelers pass on, speeding toward the Grand Canyon or Las Vegas. Perhaps they fear that "Meteor Crater" is just another tourist trap like the fake forts and dinosaur zoos that cluster along major highways in that part of the country.

The structure that lies a few miles south of the interstate is, however, one of the real wonders of the world—the best example on our planet of a relatively fresh impact crater, similar to the features that are found in such abundance on the Moon, Mercury, and Mars. The crater, hardly visible until you reach its rim, is about 4000 feet in diameter and 600 feet deep. Its rim rises about 150 feet above the flat desert. If it were located on the Moon, it would just barely be visible through a large telescope.

Meteor Crater was first seen by European settlers of the Arizona Territory in the 1880s. A prospector's report that the area was rich in iron, including pieces of pure metal that could be picked up on the

Meteor Crater in Arizona, nearly a mile in diameter, was formed 50,000 years ago when the Earth collided with an iron meteorite weighing more than a million tons. (Meteor Crater, Northern Arizona)

surface, led to the first visit by a geologist in 1891. This initial inspection was followed by chemical assays of the metal fragments, which showed them to be an alloy of iron and nickel. The most important early visitor was Grove Karl Gilbert, Chief Scientist of the United States Geological Survey. Gilbert at first thought the crater was the product of an impact, but after examining the site he changed his mind, concluding that the feature was volcanic in origin. One of the arguments he raised against impact origin was the circularity of the crater, which he thought would happen only when the projectile had fallen from directly overhead, an unlikely chance. In addition, he could find no evidence of a buried mass of meteoritic iron beneath the crater. Referring to this postulated iron mass, Gilbert wrote: "I did not find the star, because she is not there."

Over the following decades, most geologists agreed with Gilbert that this crater was volcanic, probably produced by a steam explosion. In part, they were influenced by its proximity to the San Francisco volcanic field, a few miles to the northwest, which does have many volcanic craters, including the well-known Sunset Crater, now protected in a National Monument. The volcanic hypothesis, however, offered no explanation for the fragments of meteoritic iron that surrounded the crater. Some scientists postulated a fantastic coincidence, in which a later fall of iron meteorites had taken place precisely on top of the older volcanic crater. The iron fragments on the surface, which no one could deny, suggested to others the possibility of a commercially valuable mass of meteoritic iron beneath the crater floor. This prospect attracted the geologist and mining engineer Daniel M. Barringer to the site in 1905.

Barringer was to devote the rest of his life to examination of the crater, which is still owned by the Barringer family today. Convinced that the crater was meteoritic in origin, Barringer and his associates drilled a number of test borings in the 1920s. They never found the iron they sought, but their work gradually convinced other geologists that this feature was, indeed, of impact origin. Not until the late 1950s, however, when the crater was extensively studied by the young geologist Eugene Shoemaker (more on him later), was the volcanic theory finally laid to rest.

Much of the confusion concerning the origin of Meteor Crater was actually the result of misunderstandings concerning the impact process. Neither Gilbert, who concluded the crater was volcanic, nor

Slice through an iron meteorite. (Klaus Keil)

Barringer, who was convinced of its meteoritic origin, had thought through carefully the consequences of an impact of such magnitude.

* * *

Now scientists have a much better understanding of such an event. The body that produced Meteor Crater was a mass of nickel-iron alloy with a diameter of about 60 meters (200 feet) and a mass of several million tons. At the time the impact occurred, about 50,000 years ago, this projectile had been orbiting in space for hundreds of millions of years. Eventually its orbit, nudged by the gravitational influences of the other planets, put it into a collision course with the Earth.

We do not know the speed or direction of the projectile before it encountered our planet. But whatever its initial motion, it was caught up in our planet's powerful gravitational field as it neared the Earth. The force of gravity is such that an impacting object, even if it were initially stationary with respect to the Earth, would be accelerated to a speed of 7 miles/sec (25,000 mph) by the time it struck. This speed, which is also equal to the Earth's escape velocity (the speed an object must have to escape from the Earth's gravitation), is the minimum for an incoming projectile. If the motion of the object in space were initially directed toward the Earth, then its velocity at impact would be even higher. Typically, the impact speed of meteorites is about 10 miles/sec (40,000 mph).

Although it was certainly within the capability of Gilbert or Barringer to calculate this minimum impact speed, they seem not to have done so. Associated with the motion of a body there is an energy, called its "kinetic energy." This energy of motion depends only on the speed and mass of the projectile. At a speed of 7 to 10 miles/sec, the kinetic energy of any object, whatever its composition, amounts to 100 million joules/kg (a joule is the common unit of energy, equal to about 1/4 calorie). For comparison, we may note that the chemical energy in dynamite or TNT—the energy released when this material explodes—is only about 1/25th as great (4 million joules/kg). In other words, any projectile impacting the Earth from space contains about 25 times the energy of an equivalent amount of TNT. When the projectile strikes, this energy is released almost instantaneously, and the result is an explosion. It is the explosion that generates the crater, destroying the projectile in the process. This process is very different from the common idea of a rock thumping down and plowing into the ground, which seems to have been the picture held, at least unconsciously, by Gilbert and Barringer.

You may object that this reasoning is incomplete, on the grounds that passage through the Earth's atmosphere will slow the projectile before impact. This is perfectly correct, but the effect is important only for small projectiles. If the incoming meteorite is the size of a lump of coal (typical of the meteorites displayed in museums), it will indeed be slowed by atmospheric friction and might land rather gently at the surface. For larger objects, however, air resistance becomes less important, and a mass of hundreds of tons will punch through the atmosphere as easily as a fist through a cobweb. The only exception would be an object of extreme low density and physical weakness—such as the Tunguska projectile, which gave up its kinetic energy in the atmosphere and exploded at an altitude of 20,000 feet.

Imagine the Meteor Crater projectile striking the ground at a speed of 10 miles/sec. Its kinetic energy is equivalent to about 15 million tons of TNT—15 megatons, in nuclear weapons jargon. When the mass of iron strikes, it is moving so fast that it penetrates to a depth equal to twice its diameter within just over 0.01 second. This is all the time that is required to stop its forward motion. In this instant, the energy of 15 million tons of TNT is released, turning the projectile and a part of the surrounding rock into a superheated fireball of incandescent vapor. Shock waves shatter the ground below the impact, and the blast pulverizes the surrounding rock and ejects it in an expanding cloud of debris. The effect is equivalent to the explosion at ground level of a very large nuclear bomb, except for the absence of a neutron blast or radioactive fallout. When the dust clears, a new crater is revealed, about 20 times the diameter of the original impacting mass of iron. Over the following 50,000 years Meteor Crater has been partly filled in by erosion from hundreds of thousands of rainstorms, but the essential outline of the original cavity remains.

Although this idea that a high-speed impact generates an explosion seems modern, it was actually developed in 1929 by British astronomer Forrest Ray Moulton, who was also an expert on military ordnance and had designed artillery projectiles during World War I. After years of futile effort trying to locate the mass of meteoritic iron below Meteor Crater, the Barringers called on Moulton as a consultant to estimate the size and location of the nickel-iron mass. Moulton's calculations were very disappointing to them, however, because he concluded (as we have) that the projectile was destroyed

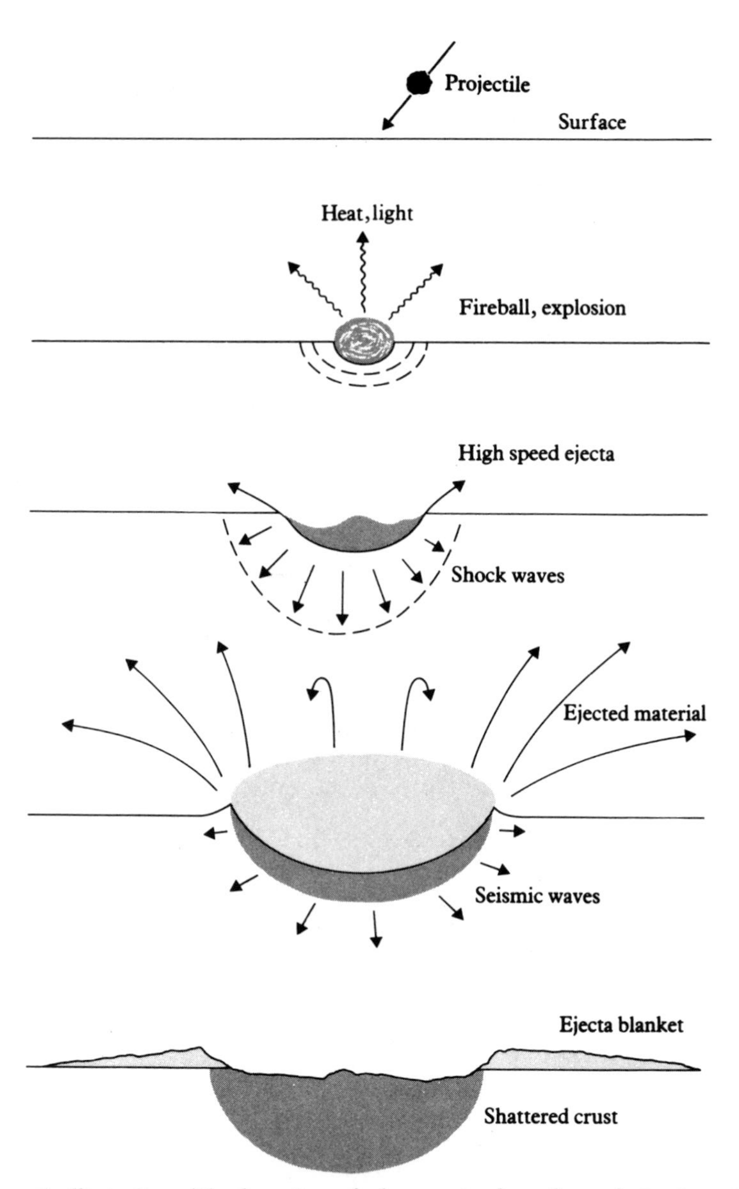

Schematic illustration of the formation of a large crater from the explosive impact of a comet or asteroid. (From *The Planetary System* by D. Morrison and T. Owen, Addison-Wesley Publishing Company, 1988)

on impact. Although these results were not published, Moulton drew upon them in 1931 when he wrote, in his popular textbook on astronomy, that "the energy given up in a tenth of a second [in a large impact] would be sufficient to vaporize both the meteorite and the material it encountered—there would be a violent explosion that would produce a circular crater, regardless of the direction of impact, which alone would remain as evidence of the event." Unfortunately, his wisdom fell on deaf ears and had to be rediscovered in the 1950s.

A quarter century after Moulton's work, Eugene Shoemaker and others of a new generation of scientists began a detailed geological examination of Meteor Crater that clearly revealed the signature of the massive explosion that created it. The rock strata at the rim of the crater are folded back like the pages of a book, so that they now lie in inverted sequence. Half-melted fragments of the original iron mass lie scattered over hundreds of square miles around the site of the impact. And the fragmented rock in the vicinity of the crater contains a number of very special minerals that can only be formed under the extreme pressures of an impact explosion: coesite, suevite, lechatelierite, and shocked grains of calcite and quartz. These mineralogical signposts remain long after most of the surface indications of the crater have eroded away, and today geologists recognize them as the surest signature of impacts that took place millions of years ago.

If Gilbert had recognized the violence of a meteoritic impact, he would not have been concerned about the circularity of the crater; all explosion craters are naturally round (look at any aerial photo of a battlefield that has been heavily bombed or shelled). The crater retains no memory of the direction from which the projectile approached; all is obliterated in the explosion. Barringer, had he realized the nature of the impact, would not have devoted 25 years to his search for the original iron mass beneath the crater, for he would have realized that no projectile could survive such an impact. Later astronomers and geologists, had they studied Moulton's text (or better yet, read his unpublished report written for the Barringers), might have been quicker to recognize the evidence for impact craters on the Earth and Moon.

* * *

No other impact feature on the Earth equals Meteor Crater in size and freshness, but other craters, both larger and smaller, are present. One of the first features to be identified as an impact crater was a shallow circular depression a couple of hundred feet across, located near Odessa, Texas. In the 1920s, Barringer's son examined the Odessa crater and concluded that "it is a meteor crater beyond the shadow of a doubt . . . . The whole thing is so absurdly like Meteor Crater that it doesn't seem possible." The Odessa crater is about 10,000 years old but is much more eroded than Meteor Crater, as a result of the greater rainfall and more frequent dust storms found in Texas. In 1937, a well-formed 500-foot crater was found near the Boxhole Sheep Station in Australia, and subsequently fragments of meteoritic iron were recovered from the site. An aerial survey conducted in Western Australia in 1947 located the Wolf Creek Crater, which is 3000 feet in diameter and almost as well preserved as Meteor Crater. Other suspected impact structures were located in Estonia, Arabia, and Africa. By midcentury, geologists recognized ten probable impact craters, in addition to Meteor Crater in Arizona.

In the meantime, other geologists had identified a new class of large structures, which they called "cryptovolcanic," each characterized by a low circular mound up to a mile or so across, with tilted and jumbled rock strata. These hills were uplifted by hundreds of feet, so that the rock layers exposed in their centers were characteristic of much deeper layers in the surrounding countryside. Although they were circular like craters, these cryptovolcanic structures were otherwise just the reverse, consisting as they did of an uplift, not a depression. Some people thought they were created by volcanic pressures from below, the unsuccessful attempt of volcanic materials to break through to the surface. But since they were not associated with known volcanoes, the mechanism of their formation remained mysterious. Some of the better-known cryptovolcanic structures are Serpent Mound in Ohio, the Kentland Disturbance in Indiana, Steinheim Basin in Germany, and Upheaval Dome in Utah. Much larger structures, such as the Bushveld Complex and the Vredefort Ring in South Africa, which are 300 miles across, also show many of the characteristics of cryptovolcanism.

During the past 20 years, as geologists have learned to identify the unique minerals and highly shocked rock fragments that are the indicators of violent explosions, these cryptovolcanic structures have become the subject of renewed research. One by one, they have revealed evidence of catastrophic formation, not the slow action of volcanic pressure. Today, most geologists are convinced that these cryptovolcanic structures are products of meteoritic impacts. Why,

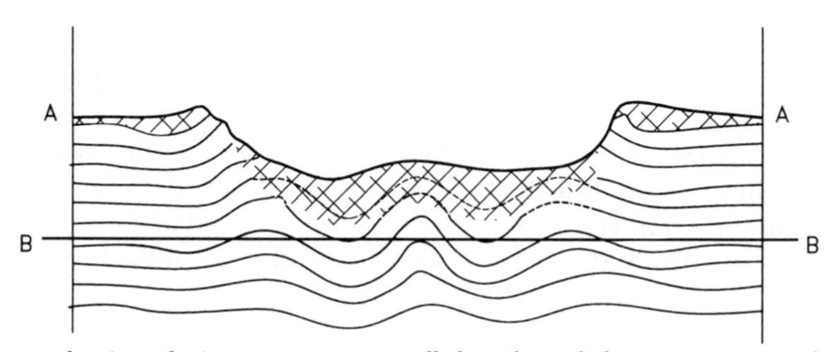

Cryptovolcanic geologic structures are actually heavily eroded impact craters, as shown in this diagram illustrating the cross section of such a feature. A shows the cross section of a fresh crater (ejecta is cross-hatched). B shows the cross section of a cryptovolcanic dome.

then, do they take the form of uplifted hills rather than depressed craters? The answer probably can be found in their advanced state of erosion (see illustration of cryptovolcanic structure). The formation of a large impact crater removes a heavy layer of overlying rock from within the crater; after the explosion is over, the material beneath the crater rebounds and warps upward. The same effect is responsible for the central peaks in many large lunar craters. This bending upward all occurs beneath the surface, so at first it is not visible. But after millions of years, erosion may remove the overburden and with it all sign of the original crater rim and cavity. What eventually remains is an uplifted circular hill made up of the rebounded rock from under the central part of the original crater.

The past few years have provided a bonanza for geologists searching for old impact craters or "astroblemes" on the Earth. Aerial photography and surveys from orbit have revealed many indistinct circular structures that could never have been recognized from the ground. Mineralogical and chemical evidence collected at these sites is now capable of identifying the signatures of violent explosions, even if these occurred tens of millions of years ago. Today more than a hundred ancient craters have been located, with diameters as great as 50 miles. The old, nearly healed scars of past impacts are thus being brought slowly to light.

* * *

Most of the eroded craters that have been discovered on the Earth are larger than Meteor Crater, and the impacts that created them were correspondingly more violent. One of the best documented of these larger impacts formed the Ries Crater 15 million years ago in the southern plains of Bavaria. The projectile in this case was a rocky asteroid with a diameter of nearly a mile and a mass of more than a billion tons—an Earth-approaching asteroid equivalent to the smallest such objects discovered today by astronomers. At a speed of 10 miles/sec, this projectile would have had an energy of several million megatons, far larger than all of the most horrible human weapons of mass destruction put together. The force of the explosion when it struck Europe is almost beyond imagining, although there have been earlier impacts that dwarf even this one.

Today the Ries Crater is a beautiful green basin about 15 miles in diameter, dotted with traditional German villages and threaded by a tributary of the Danube River. Within it lies the lovely old city of

Nördlingen, on the German "Romantic Road." Around the basin rim, however, are broken and upended rock strata, and huge boulders are found in the surrounding countryside to distances of 35 miles or more. Also present are the high-pressure minerals and shocked rocks that geologists recognize as the mark of an impact explosion site. Originally, the Ries Crater must have been at least 3 miles deep, and the mass of material removed from the cavity was more than a trillion tons.

As the mile-wide asteroid plunged through the atmosphere, it opened a cylindrical vacuum behind it, permitting some of the debris from the explosion to spray back into space. Millions of tons of this debris spread out through this hole in the atmosphere and then reentered at high speed, where friction molded the fragments into teardrop-shaped lumps of glass, typically an inch or less in size. These strange solidified droplets, called "tektites," are found today scattered over much of Central Europe. A dozen or so other tektite fields have been recognized around the world, each apparently corresponding to the melted debris from a very large impact. These curiously shaped and multicolored fragments of glass were once thought to be volcanic, while others suggested that they were formed through the fusion of atmospheric dust by lightning—another example of the reluctance of scientists to recognize the role of extraterrestrial impacts in shaping the Earth's history.

Let us try to reconstruct the effects of the Ries Crater explosion. The blast wave would have destroyed all life to a radius of hundreds of miles, while the earthquake shock must have been felt around the world. Of the trillion tons of material blasted from the crater, perhaps a billion tons of fine dust remained suspended in the atmosphere, darkening the sky all around the globe. We do not know how long this dust pall remained, or what effect it might have had on the world's ecology. If it were thick enough, it could have had long-term influence on the climate, even triggering an ice age. Fortunately, impacts of this magnitude occur only once every few million years, so we are not likely to have the opportunity to experience such an event.

The recognition that impacts have taken place on the Earth, and that they have the potential to do great damage to the ecology of our planet, is a new idea. Twenty-five years ago such concepts would have seemed absurd to most geologists, and even a decade ago their implications were only beginning to diffuse into the scientific world. But impacts are not the only unpredictable and violent events that can influence our lives. Cosmic impacts represent just one example of a general shift in thinking that has made the idea of catastrophes respectable in science, in contrast to an older uniformitarian attitude that had dominated the earth sciences for the previous 150 years. We shall discuss other such catastophic phenomena in the following chapters.

CHAPTER 3

Uniformitarianism and Catastrophism

We humans are curious about our world. Long before we had evolved the scientific method of thinking, people wondered about the forces shaping their lives. In the absence of science, our forebears—and some technologically primitive peoples today—turned to magic, or to pantheons of nature-gods, as the operative forces of Nature. In the last few centuries, civilization has developed a scientific understanding about why some things are as they are and behave as they do. But much is still to be learned. How, in the modern age, do we deal with our intense need to know, when science fails us? How do we deal with the as-yet-unknown?

In the absence of data, facts, and proven theories, we feel compelled to speculate. Few radically new theories are eventually proven to be correct, and yet can scientists say, for sure, that any particular new, wild idea is false? In the past two centuries, science has settled on an approach to deal with the dialectic between the known and the not-yet-known. It is sometimes called the "concept of uniformity," and it holds that the laws of nature (as we understand them) are constant in time: The laws operating today have done so in the past and will do so in the future. A corollary of the uniformitarian concept is that if we conduct experiments and observe nature today, we may assume that those same observed processes have acted similarly in the past. Indeed, as modern astronomers have peered into the distant past by observing galaxies billions of light years away, they have demonstrated that many physical laws were operating then just as they do today. It is, therefore, not "scientific" to invoke ad hoc catastrophes or other magical mechanisms, unknown today, to explain the past.

Uniformitarianism is useful methodology, but in the history of science—and of geology, in particular—this concept was carried to the extreme as a veritable law. Only now is 20th-century science retreating to a more balanced perspective. Two recent revolutions in the earth sciences have shown us that our world is undergoing profound, and often catastrophic, change. In the 1960s geophysicists finally recognized that the Earth's crust is divided into many separate, moving plates, which are carrying continents to and fro, and whose violent interactions are responsible for most of the mountain-building, earthquakes, and volcanism on our planet. Simultaneously, pictures sent back by far-flung spacecraft revealed solar-system-wide evidence of catastrophic bombardments and worlds in upheaval. In the wake of plate tectonics and planetary exploration, mainstream scientists have come to realize that our world is not essentially static. More and more, they are willing to entertain the notion that catastrophes may have had (or will have in the future) a dominant influence in the universe, including the history of our planet, the evolution of life, and the existence of human civilization itself.

The debate was joined nearly two centuries ago, as European geologists sought to make sense of their landscape. Speculations abounded concerning the geology of the Earth: its landforms, the stratigraphic sequence of rock layers, the varying distributions of fossils, and so on. Debates raged between Neptunists (who believed rocks precipitated from the oceans), vulcanists, and catastrophists of various stripes. In the backdrop—and this was a major factor—lay the biblical accounts of the Earth's formation and, in particular, of the Great Flood of Noah. With the Earth believed to be only 6000 years old (based on biblical genealogies), and the Flood thought to be the greatest catastrophe in the historical record, it was natural that scientists, theologians, and lay people alike would turn toward either the Creation itself or the Flood to explain most of what we see in our world.

As a young Scotsman educated in science, James Hutton traveled about Europe in the mid-18th century and began to form conclusions different from those prevalent in his day. He did not see the Earth as a completely static place, changed only by one or a few great catastrophes since its creation. Instead, his investigations revealed that it is in constant, slow change. Due to the air, the winds, the moisture, and chemical reactions, rocks are slowly decomposed and are washed,

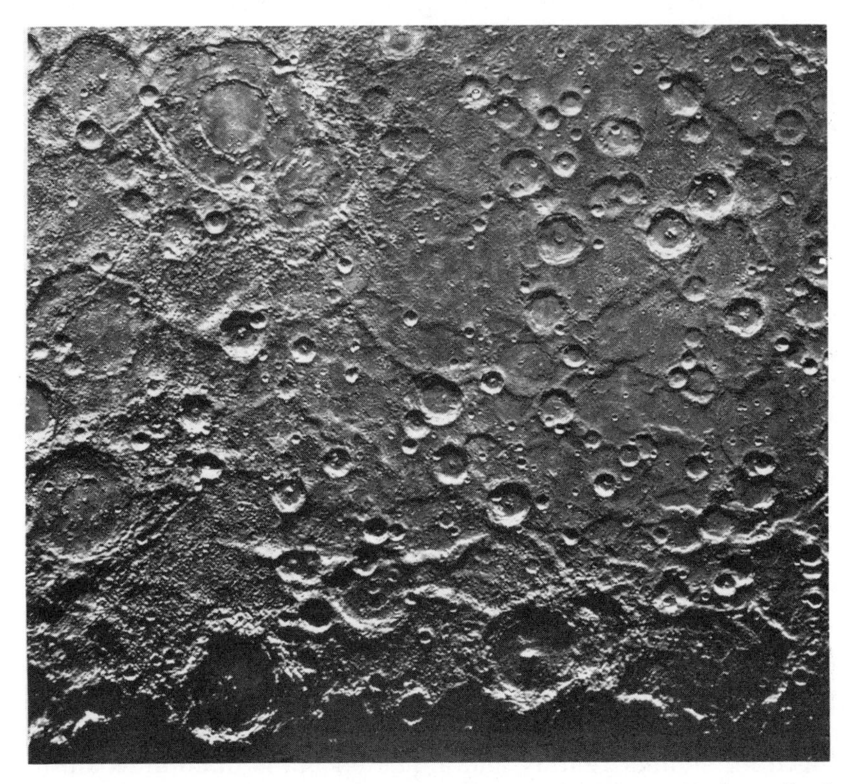

The heavily cratered surface of Mercury, as shown in this Mariner 10 photo, closely resembles that of the Moon. (NASA/Jet Propulsion Laboratory)

grain by grain, into the sea. Hutton realized that such changes are so gradual that they could alter landforms little during a human lifetime, or even after 6000 years. Only if geological history were stretched over very long durations, he reasoned, could the processes he was studying account for most geology. Hutton's concept of the calm, inexorable forces of nature, working little by little over the aeons, set the stage for later geologists to think about very slight changes accumulating over a near-eternity of time. The same concept of slow changes, when applied in the 19th century to interpreting the fossil record, would become an essential ingredient of Darwin's concept of the evolution of species.

Hutton's views presaged a debate that raged within the Geological Society in London in the 1820s between catastrophists, still wedded to the Noachian Flood, and supporters of a strict uniformitarian view promulgated by Sir Charles Lyell. Lyell, who was born the year of Hutton's death, at first studied law, but his fascination with geology led him to become the leading geologist of the Victorian age. Hutton had admitted the possibility of variations in the rates of geological processes, and catastrophists took advantage: If some variations could be admitted, why not catastrophes? Lyell, however, was strictly unyielding. As one commentator (Gillispie) wrote in 1951:

> He built his synthesis on the methodological limitation that the past could be studied only by analogy to what natural agencies can accomplish in the present. Such theoretical originality as uniformitarianism possessed lay in its pushing the analogy to an identity, in its rigorous, undeviating insistence that existing forces, given time enough, account for the observable state of man's habitat.

Lyell's opus, *Principles of Geology,* almost immediately won the argument, and geology remained in the sway of nearly strict uniformitarianism for another century and a quarter. But is uniformitarianism valid?

If you visit a creek, and observe the burbling water lapping against the banks, it is easy to appreciate the uniformitarian view. Clearly, the stream bank is slowly eroding and the mud is being washed downstream, where it ultimately reaches the sea. Imagine the stream running not just for the minutes you gaze on it, or for the years of your life, but for centuries, millennia, indeed millions of years; the cumulative erosion might well carve a Grand Canyon. But is that how the Grand Canyon was, in fact, formed?

* * *

In a desert climate, where streams aren't quite like burbling brooks, we can gain a different perspective. In the southern Arizona deserts, there are no creeks at all, except dry river channels. Obviously, minute-by-minute erosion cannot have created these channels. Yet, during the summer monsoon, occasional cloudbursts fill the rivers with turbulent, sandy waters, where adventurous teenagers risk drowning by tubing down the transient rapids for a few hours, until the rivers subside and become dry again. Perhaps it is these intermittent thunderstorms that shape the arroyo channels in the Southwest year by year, if not minute by minute.

In October 1983, something unusual happened. A dying hurricane and other chance meteorological circumstances combined to give southern Arizona as much rain in three days as it normally gets in a year. Bridges were washed away. Highways and condominiums fell into the flooding torrents. Surrounded by raging waters, Tucson became an island cut off from the rest of the world. After the flood receded, the results were obvious: Vast tracts of land had caved into the flood, and the arroyo channels had moved hundreds of feet. A thousand years of ordinary summer thunderstorms could not have caused such change! And so it is with the Grand Canyon, as well. Planetary geologist Gene Shoemaker journeyed into it to look at the places photographed by the Powell expedition in the last century. He found by comparisons that most of the changes to the canyon were the catastrophic landslides caused by rare storms, not the results of day-in, day-out erosion.

Hydrologists declared the 1983 Tucson storm to have been the "hundred-year flood"; a flood so great should be expected only once a century or so. But what about the thousand-year flood? Conceivably, an even more unlucky coincidence of weather systems (say two hurricanes at once) struck the desert some centuries ago, but the ancient Hohokam people whose ruins we now study have left no weather records. If a thousand-year flood causes more catastrophic erosion than the combined effects of ten separate hundred-year floods, then it is such very rare deluges—not the kinds of weather we witness ourselves—that are most effective in shaping the land. Just how awesome is the biggest flood in a millennium or in a hundred thousand years? Because such a calamitous storm has not been recorded recently, does that mean that it *could* not occur?

The human psyche deals best with things that are regular, orderly, constant, and moderate. We have difficulty with randomness and irregularity. We fear unexpected catastrophe. As human beings, scientists also have difficulty coming to grips with rare or unusual events, with chaotic processes. They have trouble separating reality from fiction in the case of rare phenomena. Sporadic apparitions of ghosts and flying saucers are rarely witnessed by trained observers or measured by scientific instruments. They are written off. But what is the difference between ghosts and very unusual, but documented, phenomena like "ball lightning"? Reports of these blobs of luminous plasma were dismissed as ghost stories until evidence eventually accumulated to prove that ball lightning truly exists: A few crude measurements have been made and serious theories have been formulated. Ball lightning thus has entered the sphere of science.

If a phenomenon is so exceedingly rare that it hardly ever happens, perhaps we need not care about its reality. People are killed on golf courses by everyday lightning bolts, while the much rarer ball lightning may be ignored as a mere curiosity. We must care, however, about any rare phenomenon that has enormous effects. That, of course, is essentially the definition of a catastrophe. If a single catastrophe can do more damage than the cumulative action of all the more frequent but lesser events, or than the cumulative effects of continual processes, then we must take note. To follow Sir Charles Lyell's injunction against processes that we cannot observe, or measure, in the everyday world would be a terrible blunder. As Lyell's catastrophist contemporary Adam Sedgwick complained, in his 1831 presidential address to the Geological Society, uniformitarian law

> assumes that in the laboratory of nature, no elements have ever been brought together which we ourselves have not seen combined; that no forces have been developed by their combination, of which we have not witnessed the effects.

Sedgwick rejected the view that humanity's ignorance of potential acts of God could circumscribe God's own behavior. Whether it be God's truth or scientific truth, we cannot accept Lyell's ruling out of order what might, indeed, be so.

* * *

Over the remainder of the 19th century and the beginning of the 20th, scientists tried to estimate the age of the Earth. No longer encumbered by the strict biblical account, the Earth seemed to grow older by the decade. An age of roughly 30 million years was calculated from the rate the Earth must cool since its birth, but that was before it was realized that radioactivity keeps the Earth warm. By 1900, the Earth was thought to be 100 million years old, based on the measured saltiness of the ocean and on estimated rates for sodium-bearing rocks to be washed into the sea. Geologists tried to add up the total thickness of rock layers in the ground and, from estimates of sedimentation rates, they agreed that it would take about 100 million years to lay down all the rocks. By 1930, measurements of lead isotopes in rocks (one isotope is produced from uranium, which decays radioactively at a known rate) showed that some rocks were 2 billion years old.

The real age of our planet—about 4.5 billion years—was finally accepted in the 1950s, after refinement of several radioisotopic age-dating techniques. Geologists applied the uniformitarian principle over billions of years, which is a nearly infinite stretch of time compared with the durations being considered when uniformitarianism was being developed. Until the last few decades, most geologists remained confident that the Earth operated long ago just as we know it does today. We wonder how they could have so readily ruled out even the possibilities of gross evolution of the Earth's crust, or ancient catastrophes, or changes in the physical laws as we think we understand them. Apparently, mid-20th-century geologists were imbued through their scientific education and training with traditional uniformitarianism.

Forty years ago, geologists still eschewed catastrophes, randomness, and big changes of every sort, except for the kind of modest storms, earthquakes, and volcanic explosions that were actually witnessed. Even a 1948 book on catastrophes (by the Harvard seismologist Don Leet) limits itself to earthquakes, volcanoes, tidal waves, and hurricanes; the Moon's craters might have involved some impacts, but, in the author's opinion, the falling bodies only served to trigger enormous volcanic explosions. In his view, volcanoes might spew lava, streams might meander amid the uplifting and eroding mountains, the seas might advance or recede, but the basic form of the Earth was unchanging.

Though Alfred Wegener had formulated the concept in 1910 that continents might move, geologists were still repudiating his theory 50 years later. (Wegner, who loved exploring the geology of Greenland, was trained as an astronomer and meteorologist; he was unencumbered by the traditions of geological uniformitarianism when he carried his interest in the apparent congruence of the coastlines of Africa and the Americas to its logical conclusion and founded the theory of continental drift.) The inexorable weight of geophysical data amassed by the early 1960s finally compelled geologists to reconsider our planet from a new perspective: The crust of the Earth is in continual motion due to the suite of processes now called "plate tectonics." If the geological time scale could be compressed to a short film clip, the crust of our planet would be seen to be in violent, chaotic tumult.

This intellectual revolution in geophysics brought us one new perspective; the spacecraft launched to reconnoiter the Moon and other planets yielded another. As we will describe in greater detail, the surfaces of most other planets plainly record a history of immense catastrophes. And astronomers have recently appreciated that our cosmic neighborhood is a vast shooting gallery of comets and asteroids in which the Earth itself could hardly have remained unscathed. Even to this day, some traditional geologists imagine volcanic explosions to be the mightiest catastrophes. They reject the inevitability of huge asteroidal impacts as causes for mass extinctions, such as the demise of dinosaurs 65 million years ago. At most, they acknowledge that such impacts might have triggered widespread volcanism, which is what (they believe) *really* "did in" the dinosaurs (echoing the 1948 account of lunar cratering mentioned above). Such intellectual inertia is the legacy of Lyell's uniformitarian law.

However, 25 years of planetary exploration compels most scientists to conclude that catastrophes far worse than those ever envisioned by biblical literalists must have affected, indeed dominated, our planet's early history. And future catastrophes, which dwarf even nuclear holocaust, are inevitable as well. The fact that we have not witnessed such events in our lifetimes, while welcome enough (for otherwise we wouldn't be here to contemplate them), is no reason to doubt their reality. In the heavens, we see all the asteroids and comets racing about, and by studying the planets (including our own) we can map the impact scars left by the more recent, smaller examples of the larger cataclysms that must have torn planets to their cores.

What is the difference found when comparing our emerging post-space-age catastrophism with biblically inspired catastrophism, or with its 20th-century version (worlds colliding in historical times) by the late Immanuel Velikovsky? It is fundamentally this: Our ac-

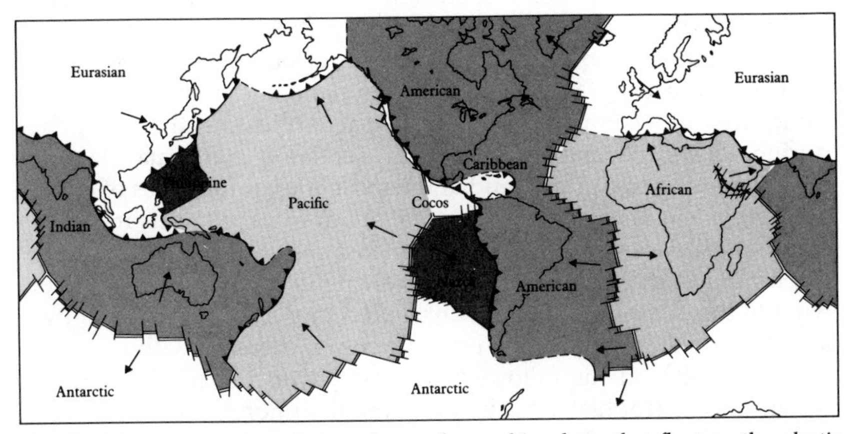

The Earth's crust consists of more than a dozen thin plates that float on the plastic mantle beneath. (From *Earth Science* by Foster, Benjamin-Cummings Publishing Company, 1982)

cumulation of scientific facts, our observations of our planet and the universe, our increasingly precise laboratory experiments, and calculations on our ever-more-capable computers have all combined to require us to believe that in some ways the universe is inherently chaotic, and that catastrophes are inevitable. The scientific process itself, not mere arm-waving or reliance on ancient texts, has led inescapably to this conclusion. Of course, our modern scientific views concern catastrophes that have very low probabilities of happening in our lifetimes; we have become more sure than ever that the great Earth-shaking cataclysms purported to have occurred during recent, historical times never actually happened.

Many knowledgeable, open-minded scientists are still wary of catastrophism of any sort, however. By their very nature, catastrophes cannot be observed today. Even our outlawed above-ground nuclear bomb tests, as "laboratory simulations" of cratering explosions, are puny compared with the impact explosions that happened during geological history. As we will see later, mathematicians are only beginning to comprehend and handle the complexities of chaotic processes. We must remain skeptical about extrapolations into times from which we have little record, or to scales beyond what we can duplicate in our laboratories, or involving complexities beyond our wit. On the other hand, we are chastened by the history of geology to avoid concluding that things we have trouble dealing with cannot be correct. Lyell and his followers overreacted to the unsubstantiated beliefs of catastrophists, but by pursuing the rigorous scientific method Lyell advocated, 20th-century scientists have been forced to regard random catastrophes as a reality. In the future, we must continue to avoid the trap of ascribing to *Nature* the simple, orderly behavior that *we* find easier to calculate and more agreeable to think about.

In seeking to deal with dilemmas still unresolved by science, some people turn to pseudoscience and check their daily horoscopes. Others search for the meaning of existence in literature and the arts. Many still turn to religion, which retreats—sometimes reluctantly—from matters now explained by science to the still-mysterious realms of human experience. Scientists, however, must deal scientifically with the as-yet-unknown, and that is difficult. Researchers cannot afford the time to analyze seriously every crackpot idea. But neither can we afford to rule hypotheses out of order simply on the grounds that we do not yet know that they are true. The wealth of new geophysical and planetary data in the last three decades has overturned many assumptions and taught us to be more humble. Hereafter, we must walk the tightrope between taking pseudoscience too seriously, on the one hand, and rejecting incompletely supported hypotheses too readily, on the other. For if catastrophism, so recently in the clutches of charlatans, can reenter the halls of science, we must be prepared to deal responsibly with whatever scientific revolution may next confront us.

CHAPTER 4

Craters on the Moon and Mars

When Galileo first turned his small telescope to the Moon in 1610, he observed a number of craters, their walls shining in the light of the slowly rising Sun. As astronomers explored the Moon and planets with early telescopes, what impressed them was not the unusualness of lunar craters, which seemed to them rather like familiar volcanoes, but that the Moon and planets looked so similar in many ways to our own world. Over ensuing centuries, people speculated about the inhabitants of these heavenly bodies. Astronomers observed different colors and forms of surface features, noted the presence or absence of clouds, and remarked on other disparities between the Moon, the planets, and Earth. But such variations were accepted as no greater than those between the deserts of Arabia and the countryside of England.

In the 19th century, several telescopic observers drew elaborate maps of the Moon. They counted the thousands of craters big enough to see, all larger than Meteor Crater, that are sprinkled across the lunar plains and are clustered in the highlands. While most astronomers still thought the craters to be volcanoes of various sorts, some alternate ideas were suggested: Could the craters be burst bubbles, or could they even be due to great meteorites plowing into the lunar surface? Despite the preponderance of scientists who were convinced they were volcanoes, the meteorite crater idea developed support in the years following World War II. By the time of the lunar landings, the impact hypothesis was in the ascendancy, where it has remained. There is widespread evidence of ancient volcanism on the Moon, to be sure, and some of the smaller craters are of volcanic origin, but the

King crater, a relatively fresh 50-mile-wide crater on the lunar far side, shows the features characteristic of undisturbed impact structures. (NASA/Johnson Space Center)

overwhelming preponderance of lunar features—including the great ringed basins that form the "man in the Moon"—are due to impact.

Ralph B. Baldwin, president of the Oliver Machinery Company in Michigan, is the man largely responsible for demonstrating that lunar craters are due to meteoritic impact explosions. Baldwin had been educated as an astrophysicist at the University of Chicago. While he was giving lectures at Chicago's Adler Planetarium in 1941, his interest was piqued by some large photographs of the Moon displayed on the walls. He didn't know about the early writings on lunar geology, including the impact-origin proposals in the 1890s by G. K. Gilbert of the United States Geological Survey. The idea that high-velocity impacts should be explosive was an unfamiliar concept to most scientists at the time, but second nature for Baldwin. He speculated that the lunar valleys radiating out from Mare Imbrium (one of the large, dark "seas" on the Moon) indicated an explosive impact origin.

During World War II, Baldwin's experiences researching the proximity fuse for bombs—for which he received citations from the Army and Navy and a Presidential Certificate of Merit—gave him an unusual new perspective on his lunar hobby. His seminal book, *The Face of the Moon*, written immediately after the end of the war and published in 1949, is replete with photographs of bomb craters in Germany, and data from the partially declassified wartime experiments on explosions. Baldwin remained a scientific loner during ensuing decades as he tended to his company and pursued lunar science as a sideline. He owes debts to few others, as he developed, single-handedly, a perspective on the Moon that still seems remarkably cogent 40 years later, two decades after the Apollo landings.

The Face of the Moon and its 1963 successor, *The Measure of the Moon*, laid out arguments for the impact origin of lunar craters in strict logical detail. Baldwin supported his thesis with extensive tables of his own measurements of crater sizes and shapes. He also drew widely on other evidence, much of it very controversial at the time, including ideas about Meteor Crater and cryptovolcanic structures. Baldwin may have done his science in his spare time, but he was no armchair speculator about heavenly processes. His books are models of rigorous research, and they were required texts for scientists who prepared the astronauts for their lunar voyages during the 1960s.

The Moon is an asphalt-colored world, covered with a powdery layer of dark, glassy volcanic soils and pebbles. Seen from afar, with binoculars, our satellite is divided into two kinds of terrain, reminiscent of Earth's continents and oceans. The brighter highlands are rugged, covered shoulder to shoulder with thousands of craters ranging up to more than a hundred miles across. The darker lowlands are called "maria" (Latin for "seas") and are overlaid by black basaltic lava flows. Some maria, which are roughly circular in shape, are the lava-flooded bottoms of immense basins formed by the biggest impact events recorded in the geological history of the Moon. The mare surfaces are flat, with just a few scattered craters to break the monotony; for each mare crater, there are about 20 highland craters. An early Soviet space probe photographed the back side of the Moon (the side we can never see from Earth), and showed that it is nearly all cratered highlands, with very few dark mare provinces.

In the early 1960s, Ranger space probes were fired into the Moon. Just before they crashed, their cameras revealed that the entire lunar surface, maria included, is saturated with millions of very small craters, the size of a football field and smaller, down to a few feet across. Craters aren't the only landforms on the Moon; there are crevices, lava channels, isolated mountains, volcanic hills, a long cliff called the "Straight Wall," and some low ridges common to volcanic plains. But these are exceptions. The lunar landscape is essentially nothing but craters, ranging from microscopic pits in grains of lunar soil to giant ringed basins many hundreds of miles across.

* * *

Ralph Baldwin was imaginative enough to follow through on the wider implications of his impact-crater theory, for our planet and others. Of the Earth, he wrote in the preface to his 1949 book:

> . . . since the Moon has always been the companion of the Earth, the history of the former is only a paraphrase of the history of the latter. The study of the Moon thus gives us a mirror throughout all time with which to study our own Earth. [Yet] the vista opened up . . . contains a disturbing factor. There is no assurance that these meteoritic impacts have all been restricted to the past. Indeed we have positive evidence that meteorites or asteroids of the requisite size still abound in space and occasionally come close to the Earth. The explosion which formed the crater Tycho on the Moon left us an interesting [structure] to study. A similar occurrence anywhere on the Earth would be a horrifying thing, almost inconceivable in its monstrosity.

Baldwin's apocalyptic vision was realistic, although decades before its time. Unlike some later lunar researchers, Baldwin thought that most of the Moon's craters were very ancient and that the bombardment rate was much lower in modern epochs, a fact not proven to everyone's satisfaction until the Apollo Moon rocks were dated (see below). But even that modest modern impact rate, as Baldwin first appreciated, is still high enough to be quite unsettling—a point to which we will return.

Ralph Baldwin also thought about the implications of his work for the other planets. In *The Face of the Moon*, he wrote:

> The Earth and Moon are not alone in space. They circle around the sun in company with eight other planets and a sky full of comets, asteroids, and meteorites. There is no reason to believe that the Earth and Moon were singled out for meteoritic bombardment in preference to the other bodies. Mars lies even closer to the belt of asteroids than does the Earth, and hence its chances for asteroidal collisions would presumably be higher. Mercury lies closest to the Sun with its great gravitational pull. It, too, should suffer numerous hits.

This all seems reasonable enough. But, with the exception of Baldwin and just a few other astronomers, there was no recognition by the scientific community, until we were well into the space age, that the other planets would be cratered, too.

The legacy of past musings about the habitation of other worlds, fueled by Percival Lowell's advocacy of a canal-building civilization on Mars, was a cultural influence that even affected scientists. (Lowell was an American astronomer who founded the observatory that bears his name; through his telescope he thought he saw a myriad of fictitious "canals" and popularized them, leading to public paranoia during the infamous 1938 Orson Welles radio show about a martian invasion.) Few astronomers made the simple back-of-the-envelope calculations that would have demonstrated the likely prospects that Mars and other planets would be found to be cratered.

When Mariner 4, the first camera-carrying spacecraft to reach another planet, flew past Mars in July 1965, it sent back a photograph of a magnificent crater, which was printed in newspapers around the world and called "picture of the century." Mariner's dozen good pictures revealed hundreds of craters on the red planet, dashing public hopes for a life-nurturing climate: Mars was just like the Moon. Most scientists were as surprised as the public. Perhaps a few craters had been expected, but initial analysis of the pictures showed *nothing but* craters, and Mars as a red-colored copy of the Moon became fixed as a scientific fact. Over the next few years, more sophisticated Mariner spacecraft were sent to Mars, revealing canyons, dry river valleys, huge volcanoes, and other important differences between Mars and the Moon. But Mars remains first and foremost a heavily cratered planet.

In 1974, Mariner 10 arrived at the Sun-baked planet Mercury. By this time, scientists responsible for interpreting the space-probe pictures expected craters, but the apparent similarity between the Moon and Mercury was striking. A theory was proposed for why the planets were all cratered, but it represented only an expansion of Ralph Baldwin's logic: The planets are cratered similarly because they are all orbiting the Sun in the midst of hypervelocity projectiles, the asteroids and comets. Evidently, by comparison with the others, Earth is unusually active geologically, so that older craters are filled in, buried, or contorted and destroyed by wind, water, and lava. On our planet alone, only the more recent impact scars remain. The other planets with solid surfaces are cratered. And, as Baldwin had already concluded, most of their craters evidently were formed early in solar-system history.

* * *

Asteroids and comets are flying around in space. And there are holes in the ground. Is there, in fact, correspondence between these known projectiles and the crater populations? Consider this hypothetical case: If all asteroids were 5 miles across, all the lunar craters they make would be roughly the same size, about 50 miles across, which is the size crater formed by the explosion of a 5-mile projectile hitting the Moon at the typical impact velocity of asteroids near the Earth–Moon system. All impact craters on the Earth would be of similar sizes, too, but larger than on the Moon (about 100 miles across) because of the acceleration of the projectiles by the Earth's gravity as they plummet into Earth's surface. The greater the impact speed, the bigger the explosion, resulting in a wider crater. (As an analogy, think about how much more damage results from a 30 mph auto crash compared with a 10 mph fender bender.)

What size craters would be made by our 5-mile asteroids impacting on Mercury and Mars? We have already noted that a target planet's gravity affects impact velocity. If everything else were equal, the craters on Mercury and Mars would be intermediate in size between those on the Moon and the Earth, because Mercury and Mars are larger than the Moon but smaller than Earth. However, there is an-

Argyre is a large impact basin on Mars, with a diameter of more than 400 miles. The mountains surrounding the basin were thrown up by the impact explosion. (NASA/Jet Propulsion Laboratory)

other factor affecting projectile velocities: The orbital velocities of asteroids and comets are faster close to the Sun, a fact encapsulated by Johannes Kepler in his famous Second Law of planetary motion. Mercury is both more massive than the Moon and closer to the Sun, so any of our hypothetical 5-mile asteroids would make craters considerably larger on that Sun-seared planet than on the Moon. However, asteroids are much more laggard in their orbits out by Mars, so the martian craters would be comparatively smaller.

Another factor determines crater size—the size of the projectile. Obviously, the larger an impacting asteroid is, the larger the resulting crater will be. And interplanetary projectiles are *not* all the same size. In fact, asteroids and comets come in a wide range of sizes. Ceres, the largest asteroid, is nearly 600 miles in diameter, while many others are about 1 mile, and innumerable undiscovered asteroids presumably exist, grading down to the boulders that wind up in our museums as meteorites. Comets also come in various sizes. In fact, the precise description of the relative numbers of small bodies of various sizes (or the relative numbers of craters of various sizes)—the "size distribution" of the population—is very important for scientists to measure. Because of the relationship between projectile sizes and crater sizes, we could compare whether or not the size distribution of asteroids corresponds to that for craters of the appropriate (proportionately larger) sizes, and decide whether or not it makes sense that the asteroids are responsible for the craters we have measured on the Moon, Mars, and Mercury. We can't tell from the Earth's craters, since the smaller ones are more likely to have been eroded away than larger ones, which skews the statistics.

The answer is not yet settled. The reason is this: Most tabulated lunar craters range between 1 and 100 miles in size (only a few craters are larger), which would be made by asteroids between several hundred feet and just 10 miles across; unfortunately, astronomers have not yet discovered most of the asteroids smaller than about 30 miles in diameter, so comparison between asteroid and crater size distributions is difficult. They seem to be similar, however, which is reassuring. But there's an important difference. The crater counts show that on both the Moon and Mars, the crater size distribution is different on the older, heavily cratered highlands than on the younger, more sparsely cratered plains. It is not simply that there are more craters on the highlands, but the augmentation in numbers of larger craters is even greater than for the smaller craters. In other words, the projec-

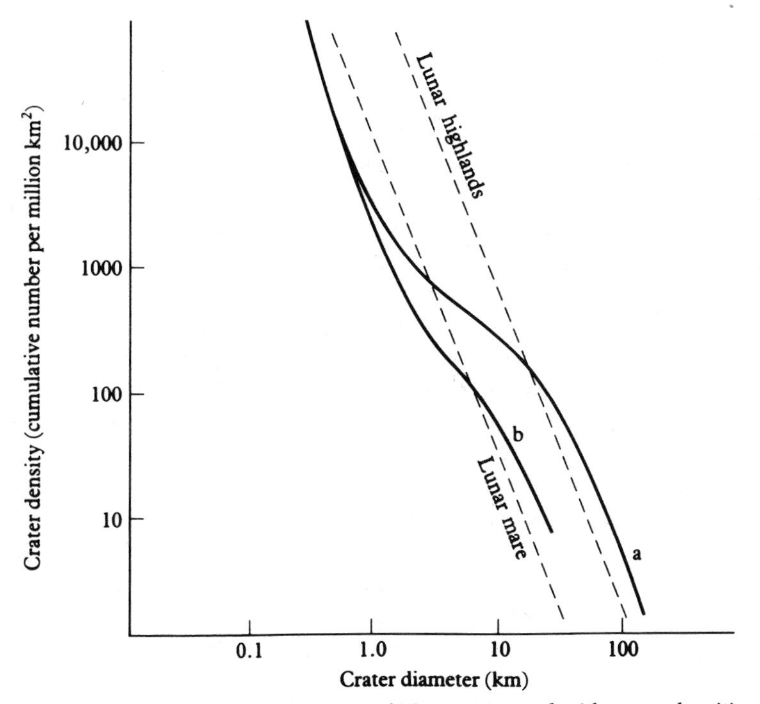

Distribution of crater sizes on two regions of Mars, compared with crater densities on the Moon. Note the much greater frequency of small craters on both the Moon and Mars. (From *The Planetary System* by D. Morrison and T. Owen, Addison-Wesley Publishing Company, 1988)

tile population responsible for the old highland craters, on both Mars and the Moon, was recognizably different from the modern asteroid/comet population responsible for the more recent cratering. It turns out that there was a distinct period of bombardment, responsible for the heavily cratered terrains. Some kind of projectiles other than the comets and asteroids that we observe today must have caused the ancient cratering impacts. As we will see, they were apparently early remnants from the formation of the planets.

* * *

The names of several scientists have appeared in our stories about cratering: G. K. Gilbert, Gene Shoemaker, Ralph Baldwin. Soon we'll have more to say about Shoemaker, but it is time to introduce a new player in this scientific saga about cratering of the Moon and planets, William K. Hartmann. Others were responsible for recognizing the importance of impact-explosion craters on the Earth and the Moon. But Hartmann was the first of the modern cratering catastrophists. An unlikely catastrophist, Bill Hartmann is a tall, lanky man of amiable demeanor. Only a part-time research scientist, Hartmann has diverse interests and abilities. He is perhaps as well known for his paintings (mostly space art) as for his science. He is also known for his writings, especially his *Cosmic Journey* textbook, a best-selling undergraduate astronomy text. He surprised his colleagues recently when a compendium appeared of science-fact essays paired with science fiction works by some of the world's famous SF writers—and it turned out that Bill Hartmann had contributed one of the fictional works. He has hosted a weekly radio show on movie music, and he has dabbled in a variety of scientific, political, philosophical, and artistic endeavors.

In his scientific work, Hartmann is individualistic, almost iconoclastic. Instead of burying himself in nitty-gritty details, Hartmann often goes straight for the Big Picture, which he often perceives before anyone else. Only after much additional data have been accumulated have Hartmann's colleagues, sometimes grudgingly, accepted the essential correctness of his views.

As a graduate student in the early 1960s, Hartmann became mesmerized by the large, round circular features on the Moon. His associates at the Lunar and Planetary Laboratory, which had just moved from Chicago to Tucson, had projected Earth-based photographs of the Moon onto a white sphere. When Hartmann stood to the side, he

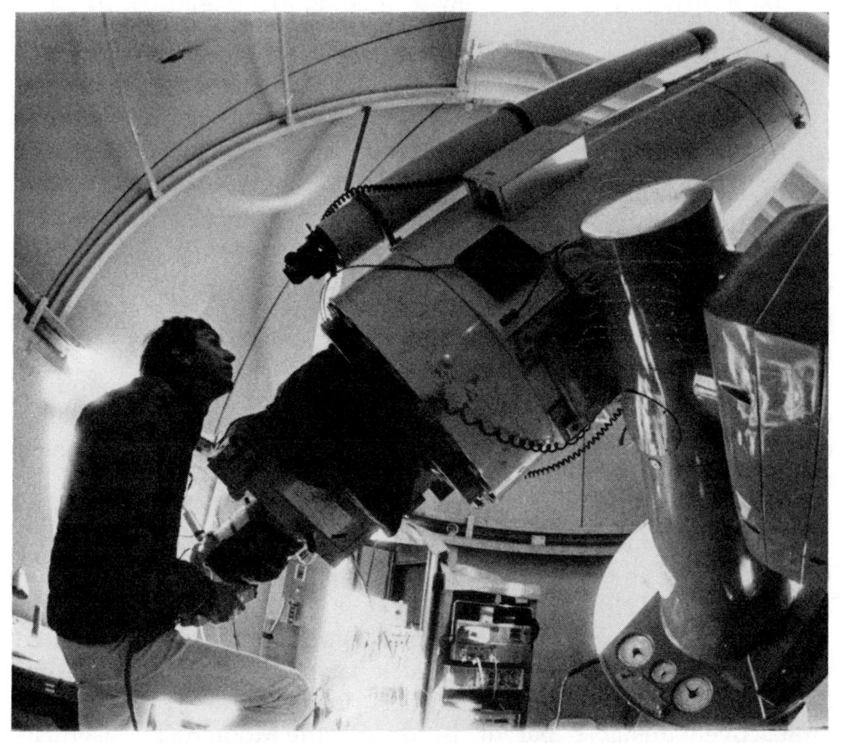

Bill Hartmann—scientist, author, and artist—was one of the first scientists to calculate cratering rates and surface ages for the terrestrial planets. (Planetary Science Institute)

could see the features near the edge of the Moon as though he were above them. He became impressed with the pervasiveness of the numerous circular basins—indeed he first applied the term "basin" to these very large craters—and he measured their associated radial and concentric features. Elaborating on Baldwin's early suggestion, Hartmann developed the case that the basins resulted from immense impacts. The next question was, "When were the basins formed?"

Bill Hartmann applied a new technique, first suggested by Gene Shoemaker, to measure the ages of the mare basins. As you might imagine, the sparsely cratered plains are younger than the heavily cratered terrains. The lunar surface is actually an "impact counter." The longer it is exposed to asteroidal bombardment, the more craters it accumulates. The lava flooding of the giant lunar impact basins, which created the maria, evidently occurred after 19/20ths of the highland craters were made. That is why there are only 1/20th as many craters on the maria. If the bombardment rate had been constant since the Moon was formed 4.5 billion years ago, then the maria would be only 225 million years old, 1/20th the age of the Moon. Bill Hartmann examined pertinent evidence, such as the numbers of asteroids and comets and the ages of craters on the Earth, and he concluded that it would have taken much longer—probably billions of years—to crater the maria. That means that the cratering rate could not have been always constant. The basins, and nearly all of the highland craters, must have formed during a bombardment blizzard early in the Moon's history. Hartmann coined the term "Early Intense Bombardment" in the mid-1960s and wrote that the maria were about 3.5 billion years old.

Hartmann's logic didn't convince many scientists, who waited until the Moon rocks were returned to laboratories for radioisotopic age-dating. By measuring the ratio of "daughter" decay products to radioactively unstable "parent" isotopes in the Moon rocks, scientists can apply the known decay rates and calculate ages for the rocks. It turned out that the Moon was, indeed, formed about 4.5 billion years ago, but most of the highland rocks were formed somewhat later and their ages were "reset" by intense shock (as in a cratering event) about 4 billion years ago. Most rocks from the lunar maria are younger, between 3 and 4 billion years old.

Belatedly, scientists agreed with Bill Hartmann's earlier chronology for lunar history. The basins were formed about 4 billion years ago during an episode renamed the "Late Heavy Bombardment"; "late" refers to the half-billion-year interval subsequent to solar system formation, but it is still very "early" by most standards. The black volcanic plains, the maria, were created during the subsequent billion years. Then the Moon cooled and "died" in a geological sense. Since then, its surface has been a passive impact counter, recording the infrequent scars of wayward comets and asteroids.

Bill Hartmann's concept was that the old lunar highland surfaces recorded the "sweeping up" of debris left over from the formation of the Moon and planets. If projectiles as large as those responsible for making the largest basins were still around half a billion years after the Moon was formed, then there must have been even more projectiles, and bigger ones, earlier on. Just such a chain of logic eventually led Hartmann to imagine that the Moon itself was the outcome of a gargantuan impact suffered by the Earth very early in its history. But we will tell that story later. For now, let's recapitulate: Most of the craters on the Moon, and on Mars and Mercury too, were formed during or before the Late Heavy Bombardment 4 billion years ago. In some sense, they were the tail-end of a half-billion-year-long process of sweeping up debris from the epochs of planetary formation. Since then, however, during the "postmare" epochs, lunar and planetary cratering has been due to asteroids and comets. The different sources for projectiles during the Late Heavy Bombardment and afterward may explain the slight but recognizable differences between size distributions of ancient and more recent craters.

* * *

Let us reconsider the crater size distributions. Forget the minor differences between the Late Heavy Bombardment projectile population and modern-day asteroids. In general, all these size distributions have a remarkable property, which explains the inherently catastrophic nature of solar-system impact history. Like the numbers of boulders, pebbles, and sand grains in a stream—but unlike the size distribution of apples on a tree, all of which are about the same size—smaller craters are *much* more numerous than larger ones. If you add up the *areas* of craters of different sizes, there is an interesting conclusion: The total area of the roughly 10,000 lunar maria craters about 1 km (0.6 mile) in size is roughly equal to the total area of the several craters on the maria that are about 100 km (62 miles) in size. And the

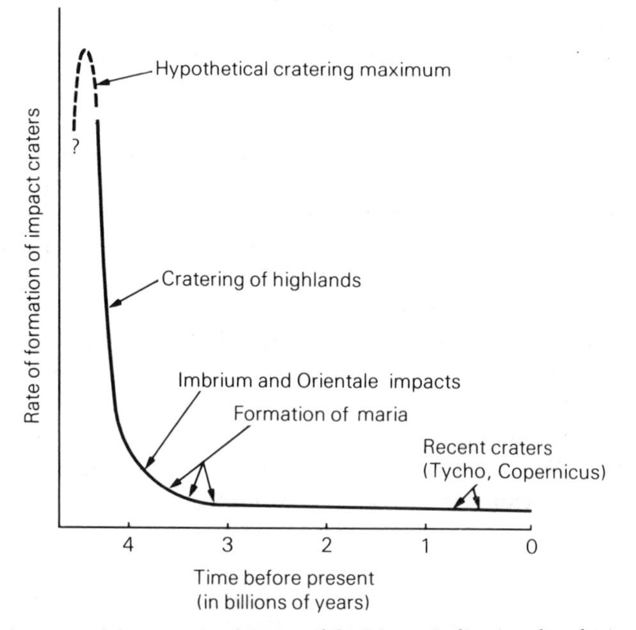

Schematic diagram of the cratering history of the Moon, indicating that the impact rates 4 billion years ago were thousands of times greater than in more recent times. (From *Realm of the Universe*, 4th edition, by G. Abell, D. Morrison, and S. Wolff, Saunders College Publishing, 1988)

total area of the 100 million craters about 10 meters (32 feet) in size is also about the same! A size distribution need not have this "equal area" trait, but this one does.

Instead of adding up crater areas, let us add up the total volumes of ejected material from the craters. Here we find that the few biggest craters contribute more ejecta volume than all the other craters combined. The volumes of the millions of tiny craters are a drop in the bucket compared with the tens of thousands of cubic miles of lunar crust expelled from a large basin impact. In other words, the same crater size distribution that is dominated in *numbers* by the smallest pits, and is balanced in *area* by all sizes of craters, is dominated in *volume* by the very biggest basin. It is just as true for the projectiles: The largest asteroid, Ceres, contains nearly as much mass as the thousands of other known asteroids put together. The energy released, and damage done, by the largest asteroidal impacts on planetary surfaces are thus more devastating than the combined effects of all of the smaller impacts put together. This is the essence of a catastrophic process: when the effects of an isolated event overwhelm the cumulative effects of everything else.

Why do the asteroids have this special, inherently catastrophic, size trait? We can philosophically regard it as a basic property of the space–time continuum in which we live. More prosaically, it results from the collisional processes, which break up asteroids into fragments, which themselves collide, yielding fragments of fragments of fragments . . . and so on. Chances of fragmentation depend on the target area presented by asteroids of various sizes to the projectile population. The bigger an asteroid, the more target area it has, and the better chance it has of getting struck, which might destroy such an area-dominating object. Thus asteroid collisional evolution tends to yield a size distribution with "equal area" traits. If a population had nearly all of the area in the biggest objects, those big ones would be struck and fragmented much more often than the smaller ones, until the areas were more in balance. In the three-dimensional universe in which we live, such equal areas imply domination, in terms of volume (or, equivalently, mass or energy), by the very biggest objects or fragments. Therefore, these asteroidal collisional processes carry with them the seeds of solar-system cratering catastrophes.

As we have seen, despite minor differences, the comets and the remnant debris from planetary formation (responsible for the Late Heavy Bombardment) have size distributions similar to that of the asteroids. Perhaps that is because the older debris were also populations of collisional fragments. In any case, the presence of such fragmental populations of projectiles whizzing around the solar system carries with it the inevitability of further catastrophes. And many catastrophes discussed in this book owe their existence to this fundamental trait of collisional processes as manifested in our three-dimensional world.

CHAPTER 5

Interplanetary Projectiles: Asteroids and Comets

Besides the Sun, the nine planets, and their moons, the solar system contains countless smaller bodies. These asteroids and comets are not very big by planetary standards. Even the largest one, the asteroid Ceres, is so small that it would take dozens of Ceres-sized bodies to make up just one body the size of our Moon. But by ordinary human standards, these small planets are very big. Ceres is nearly the size of Texas, and thousands of asteroids are larger across than the City of San Francisco. A typical icy comet body is even smaller than San Francisco—some as little as Golden Gate Park—although a comet's wispy tail of gas and dust may stretch for tens of millions of miles through the inner solar system. Such a tail streaming close to Earth can be a stunning sight in the nighttime sky, which is why so many comets have been recorded throughout history, despite their comparatively small sizes. Actually, astronomers now believe that comets are vastly more abundant than asteroids, numbering literally in the trillions. But nearly all comets are invisible: They wander far beyond the orbit of Pluto, and only dimly reflect the glow of a faraway Sun. The comets we actually see are rare stragglers, deflected in toward the inner solar system.

Because so many asteroids and comets streak around the inner solar system at speeds of many miles per second, and because they are so big by our everyday experience, we might think—incorrectly—that they would be continually colliding with each other, resulting in stupendous, blinding explosions in the nighttime sky. Indeed the asteroids are crowded together, by planetary standards, in the so-called "main belt" of asteroids, a doughnut-shaped volume of space between the orbits of Mars and Jupiter. And when they do collide and something the size of San Francisco slams at 10,000 miles per hour into a small planet the size of Ohio, the resulting devastation is enormous, beyond our comprehension. (It may be hard to believe, but such asteroid explosions are puny in comparison with other cosmic catastrophes we will discuss in later chapters.) Nearly every asteroid has suffered one or more such catastrophic collisions. Often they are cracked and shattered, sometimes destroyed. Asteroid sizes, shapes, and spins have been molded by such events.

Nevertheless, astronomers, who have been watching asteroids for nearly two centuries, have yet to witness a collisional explosion in our skies. The reason is that interplanetary space is very voluminous and very empty. The terrifying asteroid swarms depicted in science fiction movies simply don't exist in our solar system. You could be in the very middle of the asteroid belt and might be unable to see a single one. A couple of asteroids can occasionally be discerned from Earth—in dark rural skies—at the limit of vision, as very faint stars. But even from within the middle of the main belt, only a few might be dimly visible at any one time.

With such vast distances between the asteroids, even though they zoom around at many miles per second, it can be millions or even billions of years before a particular asteroid hits another one large enough to break it into pieces. Yet the solar system is very old, so there has been time enough for most asteroids to have been blown apart. Many of the smaller asteroids—those the size of Connecticut and smaller—must be fragments of preexisting, larger bodies. Even the very largest asteroids have been profoundly affected by the occasional pelting from projectiles they have encountered over the aeons. Some have been literally stripped to their iron cores.

Besides being smaller and very numerous, and coming from great distances, just how do comets differ from asteroids? The most profound difference is their chemical composition. The word "comet" comes from the Greek word for "long-haired." Like luminous tresses streaming away in the wind, comet tails have cast searchlight-like beams of light across the heavens long before people could project artificial beams to advertise movie premieres, new car dealerships, or subdivisions. The reason comets have tails is that the solid bodies (nuclei) inside their large gaseous heads are composed of ices and other "volatile" materials, substances that tend to sublimate and

Artist's impression of the violent collision of two asteroids in the main asteroid belt. (David Fischer)

evaporate when heated to room temperature. In contrast, asteroids don't form tails because their surfaces are made of rocks and metals or else they orbit quite far from the Sun, where any water-ice in their interiors remains frozen. Comet nuclei contain not only abundant water-ice and dust, but also the frozen ices of more volatile compounds—which are gaseous even at Antarctic temperatures—such as dry ice (frozen carbon dioxide), and the ices of methane and ammonia.

A typical comet nucleus has been stored in the deep freeze of interstellar space, a billion miles from the Sun, in a vast halo of comets surrounding the solar system called the "Oort cloud," named for the Dutch astronomer Jan Oort, who theorized in 1950 about the origin of comets. An occasional comet nucleus approaches the Sun, after being deflected from its path in the Oort cloud. Some, called "new comets," travel directly in toward the inner solar system; others are trapped by the powerful gravity fields of the outer planets, and wander around in the outer solar system before some of them are perturbed into short-period comet orbits in the inner solar system.

As a comet approaches the Sun, its surface ices are warmed and they evaporate. The liberated molecules are ionized—they become electrically charged—and stream away from the Sun blown by the solar wind, forming immense, luminous "ion tails." As the gases flow away from the comet's surface, they carry off the embedded dust particles, which often follow a slightly divergent path, yielding a second "dust tail."

After their swing past the Sun, new comets usually vanish back into the Oort cloud, having suffered only superficial effects from their brief encounters with solar heat. But some comets are trapped in the inner solar system as "short-period comets," and they are doomed to return to the Sun again and again, every few years. The dust that is left behind on the surface of a comet nucleus, like soot remaining on a northern city's springtime snowbanks, forms a dark crust over the ice. Gradually, after several millennia, a typical comet has made more than a thousand passes near the Sun and has lost most of its near-surface ices. It "dies": No more is there a supply of ice to generate its atmosphere and tail, and it looks like a smallish, black asteroid traveling in its elongated, cometlike orbit.

* * *

The universe was originally composed of hydrogen and helium. It took high-temperature processing within generations of stars to

Halley's Comet in 1986. Comets appear bright as a result of the extensive atmosphere and tail of gas and dust that evaporates from the small icy nucleus. (University of Hawaii photo by D. Cruikshank and A. Storrs)

Dark nucleus of Halley's Comet, a mass of ice and dust about 10 miles across, as photographed in 1986 by the Giotto spacecraft of the European Space Agency. (University of Arizona Space Imagery Center)

fabricate the heavier chemical elements of which planets, and life itself, are made. In the roughly 15 billion years since the "Big Bang" (the greatest cosmic catastrophe of all, although a uniquely beneficial one since it formed the universe we know), stellar nuclear furnaces have gradually yielded small but increasing percentages of the all-important heavier elements. The resulting "cosmic composition" of our Milky Way Galaxy is rich in original hydrogen and helium, but contains other important elements (like silicon and oxygen), and minute traces of some heavy, nonvolatile elements like uranium and iridium. Astronomers have analyzed the spectrum of the Sun, finding that it, too, contains the cosmic proportions of the elements. That same cosmic chemistry must have characterized the original solar nebula, the interstellar cloud of gas and dust that collapsed to form the Sun and the planets 4.5 billion years ago.

From what we have learned about comets, they seem to contain a representative cosmic sample of every element except for the very lightest gases (like hydrogen and helium), which are gaseous even at extremely cold temperatures and which cannot be retained by the weak gravity field of a comet. Thus, for a solid object, the nucleus of a "new" comet seems to be the most pristine remnant of the original matter from which the solar system was made, excepting only that comets lack the very most volatile gases. It has been their storage locations, near the outer periphery of the solar system, that have enabled comets to retain their primordial, ice-rich compositions.

There are many forces and processes that have operated in our planetary system to transform the cosmic chemistry with which the solar nebula was originally endowed. These processes have segregated or "fractionated" the chemicals. We have already discussed one such fractionation process: the heating, evaporation, and devolatilization a comet suffers if it spends much time at all near the Sun. Our own planet Earth is being fractionated in much the same way comets are, as the lighter gases in our atmosphere leak off into space. The fingerprints of chemical abundance patterns measured in Moon rocks, meteorites, and other extraterrestrial materials can tell us about important processes that occurred in ancient times, when the planets were forming.

The chemistry of meteorites, which are thought to be fragments of main-belt asteroids, reveals some important ancient processes. The so-called "carbonaceous meteorites" show modest departures from original cosmic abundances; they contain cosmic proportions of nonvolatile elements, but they tend to lack elements that are gaseous at room temperature, including those that comprise cometary ices. (Scientists call such relatively unaltered compositions "primitive" ones.) Both asteroids and comets formed from solid grains in the original solar nebula. Evidently carbonaceous asteroids originated closer than comets to the warmth of the forming Sun, where the most volatile icy grains were evaporated; such grains were preserved farther out in the solar system where comets originated.

Not all meteorites exhibit just the modest chemical fractionations found in comets and carbonaceous meteorites. Some appear to be solidified lava. Others are made of pure nickel-iron alloy. To understand the more extreme fractionation processes that could produce such materials, consider the familiar rocks of the Earth's crust, which are exceptional compounds by cosmic standards. One cosmically abundant material that is quite rare on the Earth's surface is iron. Human civilization has already consumed some of the most accessible concentrations of iron ores, such as the once-vast deposits in Minnesota's Mesabi Range. Seismologists and geophysicists have discovered that much of the "missing" iron is located in the Earth's core, an immense sphere of metallic iron as large as the whole planet Mercury. When the Earth was young and hot, iron coagulated and, because of its heavy weight, sank into the interior to form the core. Many other compounds that have chemical affinities for iron, such as iridium, were also depleted from the Earth's surface. But they remain in our planet to be found, if we could just dig deep enough.

Detailed chemical measurements prove that the weird, highly fractionated meteorites have nothing to do with the Earth. But evidently their asteroidal parent bodies have undergone the same kind of melting and core-formation processes that resulted in our own planet's crust and core. One of the profound mysteries about the early history of the solar system is why some asteroids got so hot that they melted while others remained relatively unaffected. In any case, the fact that we get fragments of the iron cores of small planets falling from the heavens is proof that the asteroids have been colliding and breaking each other apart. Otherwise, asteroidal iron would remain as inaccessible as the iron deep within our own planet's core.

If all the hues of reflected light are spread out into a spectrum, one can see darker lines or bands where the material in the reflecting surface preferentially absorbed the incoming light at specific colors (wavelengths). Materials of different compositions cast their own, unique spectral "fingerprints." From such absorptions in the spectra of sunlight reflected by asteroids, and from comparisons with laboratory spectra of meteorites, astronomers have learned much about the minerals of which asteroids are made. Apart from a few unusual asteroids made of lava rocks or metal, most asteroids in the inner main belt are made of silicates with flecks of metal, just like the most common meteorites in our museums. Those in the middle and outer parts of the belt are very black and have spectra indicating they are made of organic compounds and hydrated silicates, just like the carbonaceous meteorites, which have cosmic abundances of non-volatiles. Sporadic groups of asteroids located beyond the main belt have black, reddish spectra unlike any known meteorite type; they may be of even more primitive, unaltered mineralogy and could be a transition to cometary compositions. With increasing distance from the Sun, asteroid compositions seem to vary from minerals that plausibly existed in the warmer parts of the primordial solar system out to the low-temperature minerals that could exist only in the frigid outer reaches. Thus the asteroids present a tableau of the original solar nebula.

Taken together, asteroids and comets are remnants of a once much more populous collection of small bodies that were formed at the beginning of planetary history. Modern theories hold that the solar nebula, with the young Sun at its center, was gravitationally compressed into an immense disk. From it, countless small bodies coagulated: rocky asteroidlike bodies in the inner solar system, and cometlike bodies in Jupiter's zone and beyond. Over the ensuing millions of years, these "planetesimals" gradually bumped into each other and accumulated into the planets and satellites we know today. Those that were leftover in the inner solar system are the asteroids; they are most common in the main belt, beyond the orbit of Mars. Comets, the leftover icy planetesimals of the outer solar system, were stored far beyond the orbit of Pluto. Perhaps comets originated there, and were simply too spread out in the vastness of interstellar space to encounter each other and gather into a tenth planet (or an eleventh or twelfth one). Alternatively, comets may be remnants from the formation of Uranus and Neptune, subsequently flung out toward the stars by those outer planets' gravity fields. In either case, comets and asteroids—as leftover pieces from primordial epochs—constitute the source of our most precious clues about the formation of the planets. They also constitute a population of potentially dangerous projectiles, which have left their mark on the surfaces of most planets, and which threaten us with an apocalypse even today.

* * *

Although one third of the mass of all asteroids resides in Ceres alone, there are hundreds of asteroids larger than 50 miles across and also uncountable smaller ones, grading down to mere boulders and grains of sand. The numbers of asteroids of different sizes bespeak a collisional fragmentation process. You would get the same proportions of large and small fragments by smashing a brick with a hammer. A variety of evidence suggests that the asteroids have been smashing each other to bits since the beginning of solar-system history. There are groups of asteroids circling the Sun in nearly the same orbits; such "families" are thought to be groups of fragments from larger precursor bodies broken up by a speeding asteroidal projectile. Asteroid spins (rotation periods and spin-axis orientations) look as if they were caused by random collisions. As we have mentioned, some asteroids are made of iron, and they look like preexisting bodies literally stripped to their cores. Microscopic examination of meteorites, which are asteroidal fragments fallen to Earth, show evidence of high-pressure shock, just like rocks near Meteor Crater and other astroblemes, indicating that meteorites were liberated from their parent bodies by immense impacts.

One uncertainty is whether or not the asteroids have collided so much that they have literally ground each other to bits. All the asteroids put together hardly constitute the mass of the "missing planet" that generations of astronomers have believed *should* be located in the extra-wide gap between Mars and Jupiter. A number of asteroids are missing because of orbital instabilities: As we will see later when we discuss "chaos," some asteroid orbits can become wild, tossing the asteroid into the path of another planet, or flinging it out of the solar system altogether. Even accounting for these chaotic losses and asteroid collisions, some other process (perhaps due to the massive, nearby planet Jupiter) must have helped diminish the numbers of asteroids long ago. Perhaps large, speeding planetesimals flung about by the young Jupiter collided with the planetesimals that were aggregating into the would-be asteroidal planet, smashing them to bits long before they had a chance to smash each other. Today's asteroids may be the lucky few that survived an initial episode of destruction, and have also evaded subsequent catastrophic inter-

The size distribution of fragments from a brick struck with a hammer is similar to that from the collision of two asteroids in space. (Clark Chapman)

asteroidal collisions. Those survivors have told us much about the early solar system, and they remain a source for many of the projectiles that hurtle past the Earth.

* * *

At the beginning of the 19th century, when the first asteroid, Ceres, was discovered, scientists were just beginning to realize that meteorites were rocks from outer space. Previously, reports of meteorite falls had been dismissed the way we disparage reports of ghosts or flying saucers today. (The stones themselves were attributed to various unusual earthly geological or meteorological processes.) As more and more asteroids were found during ensuing decades, the idea emerged that they were fragments of an exploded planet, and that the meteorites were just smaller fragments. The asteroids were, after all, located in the large gap between Mars and Jupiter, where a planet seemed to be "missing." Although the idea of an exploded asteroidal planet permeated astronomy textbooks well into the middle of this century, most scientists gave it up long ago. It seems clear, from the traits of both asteroids and meteorites, that they must be remnants of a vast population of asteroid-sized planetesimals which somehow failed ever to form into a planet. The asteroidal planet was stillborn. Since it never came into existence, it was never there to "explode."

The idea of an exploded planet had a brief resurgence in the 1970s, however, as a Canadian physicist struggled to understand why the planets are spaced as they are. Michael Ovenden specializes in orbital dynamics, a highly mathematical and computerized discipline dealing with the motions of bodies in the heavens. Like many mathematicians before him, Ovenden had worked out an elaborate theory to explain the planetary spacings. The problem was that for his theory to come out right, it required an extraordinary thing: A Saturn-sized planet must have existed in the asteroid belt until just 16 million years ago, when it disappeared. Most scientists would have gone back to the drawing board to devise a new theory, but Ovenden took his work too seriously. He published a paper proposing that such a planet had, in fact, exploded 16 million years ago. He viewed the asteroids, meteorites, and even some comets as likely remnants of the explosion.

Had such an explosion actually occurred, it would have been far and away the most awful, stupendous cosmic catastrophe ever to occur in the solar system. To blast apart the mass of Saturn would be equivalent to the explosion of 10 million billion bombs of the size that destroyed Hiroshima. Even the fantastic collision scientists now believe the Earth suffered early in its history, giving rise to the Moon, was a thousand times smaller than the explosion hypothesized by Ovenden.

The evidence against Ovenden's explosion is overwhelming. Not even Michael Ovenden himself could imagine a possible way for such a planet to explode. We *know* why bombs explode (they are made of explosives). We even know why stars explode, but stellar forces cannot operate within planets. The asteroids themselves are in approximately circular orbits, not the chaotic elliptical orbits one would expect for explosion debris. Meteorite minerals show evidence for high-pressure shock, to be sure, but they were affected by forces like those in cratering events, not the total vaporization that would have resulted from Ovenden's explosion. Anyway, most of the events that affected meteorites have been dated as occurring billions of years ago, not 16 million years ago. Furthermore, the dispersal of a Saturn-mass of material from the asteroidal zone only 16 million years ago would have had a big effect on lunar soils, which have passively recorded meteorite impacts for aeons, but there is no sign of such an event. If a Saturn-sized planet in the asteroid belt before 16 million years ago is required to save Ovenden's theory, it is far more reasonable to abandon the theory than to abandon much of what we know about meteorites, asteroids, and solar system history.

As we have seen, there has been a recent awakening in the scientific community concerning the importance of natural catastrophes. But Ovenden's explosion that never happened is one example—and we will discuss others—where the catastrophic hypothesis is *not* the correct one. Catastrophism and uniformitarianism in the 20th century are not religions. Our beliefs about whether catastrophes have happened or not depend upon a critical, objective weighing of the experimental and observational evidence. Michael Ovenden's failures stemmed from his inability to appreciate the abundant evidence that had been amassed by astronomers and cosmochemists outside his specialty of celestial dynamics. The cosmic catastrophes that we can endorse as possibilities are those that meet rigorous scientific tests; we will continue to disparage those that fail such tests.

* * *

Asteroids are an important part of the story of cosmic catastrophes not because they are purported to have resulted from a planetary explosion, but because they themselves are projectiles that have caused havoc in the past, and may do so again, when they collide with each other and with other planets, including our own Earth. We are not concerned with those asteroids that are stored safely in the asteroid belt, but with those that are potential projectiles. Indeed, a number of smallish asteroids and comets are known to follow paths that cross the orbits of the planets. Quite a few stragglers from the inner edge of the asteroid belt are in orbits elongated sufficiently to cross the orbit of Mars on occasion, the so-called "Mars-crossing asteroids." A much smaller number of Earth-crossers are known, called the "Apollos" (the name of the god was given to the first Earth-crosser, discovered in 1932; Apollo was subsequently lost and recently rediscovered). Although they are termed asteroids, the 50 Apollos now known are believed to be a mixture of fragments derived from the main asteroid belt and of dead, devolatilized short-period comets.

Most of the scientists who observe and study the asteroids are astronomers. However, one leading asteroid observer we have already encountered is the geologist Gene Shoemaker, who as a young man proved the extraterrestrial origin of Meteor Crater. As Shoemaker extended his research to other craters, it was only natural that he became interested in the projectiles flying around in space that might be responsible for the terrestrial craters he was studying.

In the early 1960s, Shoemaker and two of his colleagues at the United States Geological Survey wrote one of the seminal papers in planetary science, entitled "On the Interplanetary Correlation of Geologic Time." Shoemaker realized that by counting how crowded together the craters are in different regions of the Moon (or on the surfaces of the Earth, Mars, or any other planet that might have them), it would be possible to date the surfaces. Obviously, a province on a planet's surface with twice as many craters as another province must be older; if the impact rate were constant, the more heavily cratered province would, in fact, be twice as old. To know just *how old* the surfaces are (in years), we need to know how many craters form per year. Only on the Earth could crater ages be measured, from the radioactive age-dating techniques that were perfected by the 1950s.

Gene Shoemaker realized that he could extrapolate the Earth's age scale to other planets, if only he could learn how the impact rate for asteroids and comets varies between the Earth, the Moon, and the other planets. To do that, he needed to look to the skies, and catalog the Apollo asteroids, the Mars-crossers, and the other planet-

Eugene Shoemaker has contributed more than any other recent scientist to our understanding of planetary cratering, in studies ranging from the geology of Meteor Crater to the search for new comets and Earth-approaching asteroids. (United States Geological Survey)

crossers. Gene Shoemaker's goal was to learn not only about the cratering rates today, but also about how they might have changed over the age of the solar system. That involved learning everything there is to know about asteroids and comets, including the celestial dynamics of their motions and how their physical character evolves.

Gene Shoemaker recently recalled that there were just eight known Apollos in the late 1950s, and astronomers had "lost" most of them during the years since they were originally discovered. During the 1960s, the energetic Shoemaker extended his studies of craters to other worlds, thanks to the new spacecraft pictures of lunar and martian craters. But in 1973, he made a deliberate career change and became a self-taught astronomer. By founding the Palomar asteroid survey, this consummate geologist was determined to do the important work the astronomers continued to neglect, to inventory Earth- and Mars-crossing asteroids in order to determine planetary cratering rates. Also, Gene Shoemaker soon became impressed with the potential importance of comets as cratering projectiles and, with his wife Carolyn, began discovering comets at a prodigious rate. (The Shoemakers have discovered more comets than has anyone else in the 20th century.)

Now nearly 50 Apollos have been discovered and cataloged. Directly or indirectly, Gene Shoemaker has been responsible for nearly all of the new Apollos, and also for most of the known Mars-crossers. He motivated a number of asteroid searchers at Palomar Mountain Observatory, one of whom—Eleanor Helin—has, in turn, inspired many others around the world to photograph the skies and search for the faint, rapidly moving, trailed "star" images that signify a nearby asteroid. After countless cold nights (many of them cloudy ones, as all astronomers have learned to accept), Shoemaker can now claim with considerable confidence that there must be about 1100 asteroids larger than 1 km (0.6 mile) in size whose orbits can cross the orbit of the Earth.

An asteroid or comet in an orbit that crosses the orbit of a big planet is in a precarious situation. Eventually, such an object is likely to collide with the planet, or else pass close enough that the extra gravitational "kick" would shoot the body out of the solar system altogether. On average, an Apollo asteroid lasts in the inner solar system for several hundred million years. As some are lost, others replenish the supply. Some come from the main asteroid belt, due to jovian forces combined with the newly recognized propensity for some orbits to become "chaotic." Other Earth-approachers, perhaps a third of the total, are dead comets that have evaporated away their surface volatiles, as they do after a thousand or so passes near the Sun.

A thousand mountain-sized objects flying around in space, each capable of colliding with the Earth at 10 miles per second, might seem to be a terrifying threat. Indeed, as we will discuss later, there is the possibility that one of the many hundreds of Earth-crossers Gene Shoemaker has *not* yet discovered may cause Armageddon at any time. But space is voluminous and our planet a very small target, so the chances are that our next encounter with a mile-sized object won't occur for hundreds of thousands of years. In this cosmic crapshoot, one can never be certain, however. Of more general concern are the innumerable smaller objects (e.g., the size of a supertanker) that must exist, one of which stands at least a slim chance of impacting the Earth within our lifetimes. If such a body were to hit in Nevada, it would be a headline-making event. It would be much more serious if it were to plummet into Central Park in New York City, wiping out much of Manhattan, or if its flaming crash to Earth were somehow mistaken for a nuclear attack.

The numbers of larger asteroids and comets can now be securely estimated, thanks to Shoemaker's telescopic program. We have a good idea about the "interplanetary correlation of geologic time," at least in the inner solar system during the last few billion years. One of the incontrovertible conclusions is that objects as large as 10 km (6 miles) in diameter must run into our planet every 50 or 100 million years or so. As we shall see, just such an event may have profoundly altered the course of life on our planet, and it may even have been responsible for enabling human beings to evolve.

CHAPTER 6

Death of the Dinosaurs

If a small comet or asteroid should strike the Earth tomorrow, the damage could be very great or relatively small, depending on where the projectile struck. There still remain unpopulated areas of our planet, and the explosive formation of even a 5-mile-wide crater in such a place (Western Australia, the Gobi Desert, the Sahara, or even central Nevada) would not dramatically affect most of the people in the world. On the other hand, such an impact in a heavily populated urban region would kill millions of people, dwarfing the effects of the Hiroshima or Nagasaki atom bomb. But what of still larger projectiles? At what point would the catastrophe become truly global, regardless of the exact point of impact? Our planet is a big place, but worldwide catastrophes caused by collisions are by no means impossible.

Consider the effects of a collision with an asteroid or comet having hundreds of times the mass of the Ries Crater projectile discussed earlier. Such a projectile, with a diameter of about 10 kilometers (6 miles), is about the size of bodies that formed the prominent lunar craters Tycho and Copernicus, both of which are easily seen on the Moon with a small telescope. A trillion tons of material streaking toward the Earth contains a prodigious quantity of energy, equivalent to 100 million megatons—the explosive power of 5 billion atom bombs of the Hiroshima type. The resulting crater would be over 100 miles across; its initial depth might be sufficient to blast through the crust and penetrate into the upper mantle of the Earth. Calculations show that about 100 trillion tons of pulverized rock would be lofted into the atmosphere, consisting of the material of the impacting body mixed with a much greater quantity of rock dust from the terrestrial crust. If the impact occurred in the ocean, towering tidal waves would spread out over much of the planet, and additional trillions of tons of water would be vaporized.

From the lunar cratering record as tied down by age-dating of Apollo Moon rocks, we now know that one or two craters as large as Tycho or Copernicus are formed on the Moon every billion years at current bombardment rates. Can we use this information to estimate how often a 10-km projectile collides with the Earth? We certainly can, since Earth and Moon are subject to the same bombardment by interplanetary debris. The chief difference between the two is the area of the target. Of course, the Earth is considerably larger than the Moon, so we should expect such an event perhaps once every 50 million years or so. Since these are chance events, the times cannot be predicted, and the individual impacts should be randomly spaced in time. However, on the average such events should be reasonably common on the astronomical—or geological—time scale.

Although from this perspective the impact of a 10-km asteroid or comet seems inevitable, and even routine, the idea is actually a radical one for traditionally trained geologists. Before the 1980s, one would find no discussion in geology texts of the possibility of such an explosion. Similarly, no one seems to have considered the effect of such a global catastrophe on living things. This attitude began to change with the 1980 publication in *Science* magazine of a historic paper entitled "Extraterrestrial Cause for the Cretaceous–Tertiary Extinction."

* * *

Luis W. Alvarez was a Nobel-Prize-winning physicist and, until his death in 1988, was Professor Emeritus at the University of California at Berkeley. His son Walter Alvarez is a well-known Professor of Geology at Berkeley. In the late 1970s they began a collaboration to study the concentrations of trace elements in ancient rocks, in order to find clues that might help them determine the rates at which sediments were being deposited millions of years ago. The elements they selected are the so-called "siderophile," or "iron-loving," elements, which are exceedingly rare in the Earth's crust because they, like iron, are mostly segregated into the Earth's core. The expected concentrations of such elements as iridium and platinum in most rocks are a few parts per trillion, which is very small indeed. For comparison, one person in a population of 5 billion represents 200 parts per trillion of the Earth's human population. Only recently have scientists developed reliable techniques for measuring such small chemical concentrations.

Of particular interest to Walter Alvarez and other geologists was the layer of sediment that marks the transition, 65 million years ago, between the Cretaceous and Tertiary periods of geological history. The Cretaceous is the last period of the Mesozoic Era, the Age of Dinosaurs, while the Tertiary is the first period of the modern, or Cenozoic, Era. This transition represents one of the major turning points in the development of life, when the dinosaurs perished, to be replaced by a growing diversity of mammals. The cause of the extinction of the dinosaurs has been a matter of speculation and dispute for decades, with suggestions offered by paleontologists that vary from major climate changes to epidemics to the eating of dinosaur eggs by early mammals. All of these mechanisms were consistent with a uniformitarian philosophy, and most paleontologists thought that the dinosaur extinction took place over a long time span, perhaps a million years.

The dinosaurs were fascinating creatures, but there were never many of them, with only about 150 species identified during the 140 million years they existed. Consequently, their fossil bones are fairly rare—sufficiently so that the time of their extinction cannot be located precisely in the geological record. The exact boundary between the Cretaceous and the Tertiary, however, is defined by an extinction equally dramatic, and much more easily observed, than the end of the dinosaurs. At a well-determined point in the layers of sediment dating from that time, there is a sudden change in the fossil population of small marine organisms, such as the animals called "forams," which have tiny shells not much larger than a grain of sand. In the lower (older) rock there are countless shells representing thousands of different species of forams. Just above the boundary, most of these species are absent—extinct—and over time a population consisting of newly evolved species of forams can be seen to have developed. This is an example of a great dying or "mass extinction," one of the most remarkable events in the history of life on our planet.

At the precise boundary where this great dying took place, the rocks of the Cretaceous and Tertiary periods are separated by a thin layer of clay that has been found all over the world (wherever it has not been destroyed by subsequent erosion). The Alvarez team did their initial work at a prominent exposure of this boundary near the

A thin, bright layer of clay less than an inch wide (toward the end of the rock-hammer handle, separated from the thick bright sandstone by a narrow seam of coal) marks the debris from the catastrophic event that ended the Cretaceous era 65 million years ago. Here the boundary is shown in an outcrop near Madrid, Colorado. (Photo by Alan Hildebrand)

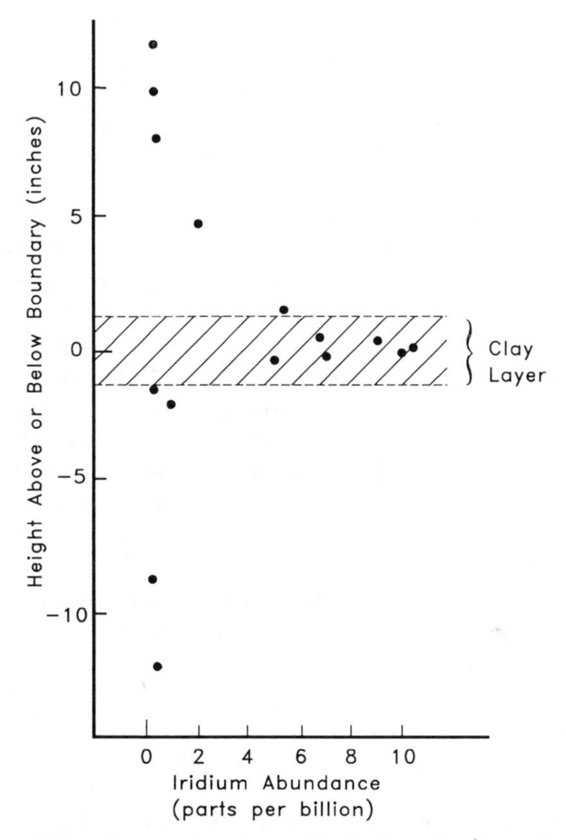

The rare element iridium is strongly enhanced in the boundary layer that marks the end of the Cretaceous, pointing toward an extraterrestrial origin for this worldwide catastrophe.

medieval town of Gubbio in northern Italy, where the clay layer is about 2 inches thick. To their surprise, they found that the concentration of the rare metal iridium in the clay was nearly 100 parts per billion, or 300 times greater than in the rocks on either side of the boundary, or in other rocks from both the Cretaceous and Tertiary periods. Where did this anomalous iridium come from? The Alvarezes knew that most of the iridium in terrestrial sediments is meteoritic in origin. This is so because there is so little natural iridium in crustal rocks, while unfractionated, primitive debris from space contain a proportionately much higher concentration of this and other siderophile elements. But what could cause a concentration of meteoritic material precisely at the time of the great dying?

Luis and Walter Alvarez made the bold suggestion that a sudden influx of meteoritic material into the atmosphere was responsible for both the iridium-enriched clay and the simultaneous mass extinction. Since the clay deposit is global in extent, the amount of material it contains is simply equal to the observed thickness times the surface area of the Earth, or about 100 trillion tons. From the measured concentration of iridium, one can also estimate the mass of injected meteoritic material, which is about 1 trillion tons—equivalent to a projectile 10 km (about 6 miles) in diameter composed of primitive material. Such a body could have been either an asteroid or a comet, although a comet would have to have been somewhat larger since cometary ices are devoid of iridium.

Having made this suggestion to explain the iridium anomaly, the Alvarezes recognized that they were also onto a plausible explanation for the mass extinction. The explosion produced by such a large impact would itself be damaging, but even more important would be the consequences of lofting 100 trillion tons of dust into the atmosphere. Scientists have calculated that this quantity of dust would have produced a pall over the whole globe with a duration of at least several months. During this period of global darkness, neither light nor heat was available to sustain life. With photosynthesis at a standstill and the Earth plunged into deep cold, it is no wonder there was a great dying.

* * *

The Alvarez proposal that the mass extinction at the end of the Cretaceous was the result of an extraterrestrial impact caught most of the scientific community by surprise. Catastrophist explanations for mass extinctions had not been given much credence by paleontologists, certainly not catastrophes from the heavens. Few planetary scientists had thought much about the implications for the Earth's climate and environment of the cratering record on the Moon and Mars (although Ralph Baldwin, Eugene Shoemaker, and others had mentioned some of the possibilities). While some scientists welcomed the 1980 paper as a breakthrough, others raised objections. One leading paleontologist, David Raup of the University of Chicago, expressed both perspectives in confidential comments (since published) that he provided to the editors of *Science* evaluating the original Alvarez manuscript. Raup began his review as follows:

> The potential impact of this paper has cosmic proportions. If the hypothesis is correct, it will have profound influence on geology and evolution—not to mention philosophy. The paper will probably change the basic thinking of many people. This is fine if the hypothesis is correct. But if it turns out (later) to be wrong, a lot of damage will have been done and it will take years to recover. Because the hypothesis of the paper goes strongly against conventional wisdom, many readers will do their utmost to find fault.*

In proper scientific fashion, critics tried to find contradictions that would falsify, or disprove, the impact hypothesis. Any major contradiction is usually sufficient to relegate a new and unorthodox hypothesis to the scientific trash bin. One of the most serious of these objections is the absence of any 100-mile-diameter crater on the Earth with an age of 65 million years. Without a corpse, it is difficult to establish that a murder has taken place. While no such crater exists on the land area of the Earth, we have insufficient knowledge of the ocean floor to be sure there is no buried crater matching this description under the sea. Also, the active plate motion of the Earth's crust is constantly creating new ocean floor at rift zones, while older parts of the floor are destroyed. Over the past 65 million years, more than a quarter of the ocean floor has been recycled by plate tectonics, so there is a reasonable chance that an original crater on the ocean floor would have been completely destroyed.

Many scientists became interested in the theory after the 1980 publication of the Alvarez paper, and they set out to test its implications. In 1987, the year before his death, Luis Alvarez published in *Physics Today* a summary of more than a dozen such tests, all of which support the impact hypothesis. We'll examine some of the results, adapted from his list.

*D. Raup, *The Nemesis Affair*, Norton (1986).

The original discovery of an iridium anomaly was confined to a single location: the Gubbio clay layer, which was formed 65 million years ago by submarine deposition. From this single measurement result, the Alvarezes suggested that the impact had produced a worldwide catastrophe. If their hypothesis is correct, then the fallout from the global dust, appearing now in the form of the boundary clay, should also be worldwide, and not confined to northern Italy. This fallout should appear in deposits that formed on land as well as beneath the ocean. Further, they predicted that the iridium anomaly should be worldwide, with approximately the same concentrations of this element in the boundary clay wherever it is measured. Finally, they suggested that the clay had been deposited relatively rapidly, as the globe-circling dust settled out of the atmosphere. It follows that the layer should have about the same overall chemical composition everywhere, and that this composition should be distinctly different from that of the sediments above and below, which were the result of different processes.

Many geologists set out to test these predictions by collecting material from the Cretaceous–Tertiary boundary at sites all over the world. Unfortunately, most of these locations are not directly accessible near the surface, and indeed over much of the Earth subsequent erosional and other geological activity has destroyed this part of the geological record entirely. Soon the data began to come in, however, and they supported the impact hypothesis. One of the most important early findings was from the Raton Basin in New Mexico, where the boundary clay was located in a drill hole at a depth of 800 feet. At this site the iridium-enhanced deposit was observed in continental rocks formed on dry land. This discovery established that the boundary material had been deposited from the atmosphere, and that it was not the result of some special process peculiar to the ocean environment, as had been suggested by some critics. By 1987 there were 75 examples in the scientific literature where iridium anomalies had been measured in the boundary clay, all occurring in the exact same age stratum as far as could be established by other measurements.

The next step was to compare the chemical compositions of the boundary clay collected at the different sites. Before the impact hypothesis was proposed, most geologists had assumed that the clay in the boundary layers had the same general composition as other clays deposited in submarine environments, in both the Cretaceous and Tertiary periods. Such was not the case, however. The proportions of elements in the boundary clay were found to be different from those in other clays located above and below the boundary. Even more striking, the compositions of boundary clays taken from sites in different parts of the world agreed with each other. In addition, it has been found that the relative proportions of isotopes of the siderophile elements in the clay—just those elements that are hypothesized to have an extraterrestrial origin—match the abundances in the meteorites.

Still another prediction of the impact hypothesis is that similar iridium enhancements in the geological column should be rare. As we have already noted, an impact of a 10-km asteroid or comet is expected only about once every 50 million years, judging by the cratering record of the Moon. If iridium enhancements are truly due to cosmic impacts, they should be comparably rare, while if more ordinary terrestrial processes are responsible for such enhancements, they might be much more common. So far this prediction seems to be confirmed. While additional iridium enhancements have been indicated at the locations of a few other mass extinctions, nothing else has been discovered of the magnitude of the Cretaceous–Tertiary anomaly. Most of the sediments that make up the miles-thick geological column display very low concentrations of siderophile elements. In fact, the general level of these elements in sediments is just consistent with the steady-state accumulation of meteoritic materials from the tons of ordinary meteors and meteorites that strike the Earth every day. However, we should add that only a small percentage of the total geological column has been checked for iridium, so the final word is not yet in on this prediction of the impact hypothesis.

So far the data we have discussed support the Alvarez hypothesis, but alternative explanations for the iridium-enhanced clay layer are possible. Many geologists are familiar with volcanic eruptions, and it is natural to ask if a giant volcanic explosion might also produce a global cloud of dust, perhaps even one that contained 100 trillion tons of material. Such an alternative does not explain the iridium enhancement, of course, but conceivably there is some special kind of rare volcanic event that might eject iridium and other siderophiles into the atmosphere. In any case, it is worthwhile to examine the implications of this alternative suggestion.

Although they are certainly violent by ordinary standards, volcanic explosions (such as those at Krakatoa in 1883 and at Mt. St. Helens in 1980) are puny in comparison to the large impact events we have examined. Both the temperatures in the fireball and the pressures exerted in the explosion are much higher in the case of an asteroidal impact than in even the most awesome volcanic explosion. We have noted that high-pressure minerals like coesite or shocked quartz provide one of the most important signatures of impact processes. The impact hypothesis predicts that similar minerals should be found in the worldwide boundary deposit, while the existence of these high-pressure and high-temperature minerals would falsify the volcanic hypothesis.

The record supports the impact hypothesis. All over the world, the boundary layer contains both shocked quartz and sand-size spherules of sanidine, a glasslike mineral formed only at high temperatures. These discoveries, announced in 1984 and 1981, respectively, effectively eliminated the volcanic alternative as far as most scientists are concerned. In late 1988, at a "global catastrophes" meeting held at the Snowbird ski resort in Utah, Arizona State University researchers announced their detection of stishovite at the Cretaceous–Tertiary boundary; experts believe that this high pressure mineral can be formed only by impact. An additional argument for a big impact comes from the global presence of the sand-sized grains. These particles are too large to be suspended in the atmosphere and carried around the globe by ordinary winds. Their presence in so many different locations seems to require a massive explosion that rips away part of the atmosphere and showers debris all over the planet. No volcanic explosion could do that, even if it somehow contrived to make the shocked quartz or sanidine grains in the first place.

Another class of implications of the impact hypothesis concerns the nature of the great dying at the end of the Cretaceous period. The boundary had originally been defined in terms of the mass extinction of forams and other ocean creatures, since these are so common in marine sediments. The impact hypothesis, however, predicts that land plants and animals should also be killed in large numbers at the time the boundary layer was deposited. At the Raton site in New Mexico, where the boundary layer was formed on land, there is evidence of just this effect. At the exact location of the iridium enhancement there exists a dramatic drop in the count of pollen grains fossilized in the rock. Evidently the pollen-producing plants in that part of the world were pretty well eliminated by the effects of the impact.

The evidence cited here provides strong support for the Alvarez hypothesis of an impact at the Cretaceous–Tertiary boundary, which generated a global dust pall and was responsible for the great dying that took place on both land and sea. The alternative, of a huge volcanic explosion, seems not to be viable, although it still has its proponents. But we tend to agree with Luis Alvarez, who wrote:

> It is now time . . . to recognize that [the impact theory] is the only existing theory that agrees with all of the observations. I feel that the shoe is now on the other foot, and that those who are pushing for the acceptance of [nonimpact] theories . . . should tell how they can overcome the many objections [described above].

* * *

Let us grant that a major impact occurred on the Earth 65 million years ago, and that it caused the mass extinction that marks the end of the Cretaceous period of the Mesozoic era. Does it also follow that this catastrophic event was responsible for the demise of the dinosaurs, which had thrived for more than 100 million years? Certainly that would seem to be the most straightforward conclusion, given that the end of the Mesozoic has always been defined by the passing of the dinosaurs. Yet it is interesting that this conclusion has been challenged by some paleontologists.

The problem is that there are relatively few dinosaur fossils, which makes the determination of the exact date of their death rather difficult. When we visit natural history museums and see rooms full of reconstructed dinosaur skeletons, most of us tend to forget just how rarely a skeleton is preserved in this way. It is likely, for example, that if every human on Earth should die tomorrow, few if any would end up as fossils. Similarly, if all of the dinosaurs alive at the time of the great impact had been killed instantly, it would be surprising if we had found even one of their fossils embedded at the boundary layer. Only in extremely fortuitous circumstances is a skeleton of a large land animal buried and transformed into stone (fossilized) rather than left to decay into the soil. In contrast, fossilized shells, composed of calcium carbonate mineral, are readily produced from the remains of the vastly larger numbers of tiny marine creatures.

There are few places on Earth where the geological record of dinosaurs and other land animals is preserved right up to the iridium-enriched clay that bounds the Cretaceous period. Only in such locations can paleontologists actually measure the extent to which the dinosaurs were still flourishing at the time of the impact, and also estimate the suddenness of their demise. One such location is at the

Fort Peck Reservoir in Montana. Here, the most recent dinosaur skeleton was found about 7 feet below the boundary layer, a distance that represents thousands of years of time. However, over the entire area studied, the average vertical spacing of dinosaur fossils is about 3 feet, corresponding to one fossil formed every few thousand years. From such sparse data, one simply cannot be sure of the extent to which the dinosaurs were wiped out by the impact itself, although the circumstantial evidence suggests this was the case.

What the fossil record does tell us is that dinosaurs flourished for tens of millions of years preceding the great impact. Perhaps they were on the decline by 65 million years ago, and perhaps not; this question is in dispute. But there is no dispute that two dozen of the dinosaur species, both herbivores and carnivores, living on the land, in the sea, and in the air, were still in existence up to essentially the time of the impact, while none at all appear in rocks formed after the event. It seems to us that it strains credulity to imagine that the demise of these great animals was unrelated to either the impact or the well-documented mass extinction of marine life that is so clearly marked at this time.

How could the impact of a 10-km asteroid or comet kill such a large fraction of the Earth's creatures, including the dinosaurs? No one is sure of the details, but a number of suggestions have been put forth. To start with, there is the global cloud of impact-generated dust, sufficient to produce a layer an inch or more thick over the whole planet when it settled back to the ground. Calculations show that such a quantity of dust, when suspended in the atmosphere, could block all sunlight from reaching the surface. Temperatures would have dropped, photosynthesis ceased, and perhaps 99% of the individual organisms on our planet perished within a few weeks. What happened to the other 1%, including many of the dinosaurs, would depend on how long the dust pall remained. Unfortunately, atmospheric physicists are unable to calculate this lifetime with confidence, and estimates vary from a few weeks to more than a year. It is also unclear whether the climate recovered relatively quickly, or whether an extended ice age might have been triggered by the transient dust cloud.

Other effects of the impact may also have been important. The blast wave itself, together with heating of the atmosphere by the fireball, could have done a great deal of damage. If the impact occurred in the ocean, as seems likely, huge tidal waves would have disturbed seafloors and devastated much of the adjoining land areas. Indeed, evidence of tidal-wave deposits has been found recently at several Cretaceous–Tertiary sites around Texas. Calculations at the Massachusetts Institute of Technology suggest that the blast wave would have initiated chemical reactions in the atmosphere, producing large quantities of nitric acid. The resulting acid rain—consisting of nearly pure nitric acid—might have persisted for weeks and poisoned lakes and oceans, with lethal effects on both plant and animal life. Finally, it has even been suggested—on the basis of soot found in boundary deposits—that there was a global firestorm that burned all of the world's forests.

Some living things must have survived this devastation, or we would not be here to write this book, and you would not be reading it. In general, conditions seem to have favored survival for creatures with relatively small size, large populations, and global distributions; also, deep sea creatures were favored over those that lived on land. Other lucky species were those that propagate by creating spores, which can lie dormant until normal environmental conditions return. Some of the survivors were small mammals—the ratlike creatures that were our ancestors—although many genera and species of mammals also became extinct as a result of the impact. When the Earth recovered from the catastrophe, the surviving mammals, like many other previously inconspicuous life forms, multiplied and evolved rapidly to fill the ecological niches that were vacant as a result of the great killing. In this way the impact that occurred 65 million years ago fundamentally altered the course of biological evolution, initiating the chain of events that ultimately led to us.

CHAPTER 7

Death Stars and Comet Showers

The Moon's cratered face provides ample evidence for past violent impacts in our part of the solar system, and the Cretaceous–Tertiary boundary layer is a tangible example of the effects of one such impact on the Earth. Surely this well-studied terrestrial catastrophe is not unique. Even if the Cretaceous–Tertiary event were the most recent major impact on the Earth, similar collisions with comets or asteroids, and perhaps some that were even more violent, must have taken place previously. What does the geological record tell us of other impact catastrophes?

To a paleontologist or geologist, the mass extinction at the end of the Cretaceous is one of a number of such discontinuities observed in the geological column. Indeed, the very definitions of many geological boundaries are based on populations of fossils (especially the remains of small marine organisms). Generally, these boundaries coincide with mass extinctions in the fossil record, which provide convenient worldwide markers to distinguish one chapter in geological history from another.

A dozen or more mass extinctions are evident in the record of the past 570 million years, which is the time period during which there are sufficient marine fossils to make such record-keeping meaningful. (Previous to 570 million years ago, few life forms had the hard shells or skeletons needed to leave prominent fossil remains.) Not all mass extinctions are equal, but four major ones are known in addition to the event that ended the Cretaceous period 65 million years ago. The greatest of these came 250 million years ago, at the end of the Paleozoic era, when it is estimated that three-quarters of all genera (and more than 95% of the individual species) in the Earth's oceans were destroyed. The event at the end of the Cretaceous resulted in the extinction of 40% of marine genera. The other three major events were 215 million years ago at the end of the Triassic period, 360 million years ago at the end of the Devonian period, and 435 million years ago at the end of the Ordovician period.

Today, about 2 million species of plants and animals have been identified on the Earth. Large as this number is, it represents only a small fraction of all of the species that have lived in the past, since the first documented life-forms of 3.6 billion years ago. In fact, it is estimated that for every species in existence today, there were several hundred others that thrived at some time in the past and then became extinct. Even at higher levels in the hierarchy of life, most genera and families from the past have also perished.

At any given time, some species are dying out, while new ones are evolving. This is by no means a continuous process, however. Following the lead of Stephen Jay Gould of Harvard University and others, most students of the history of life today recognize that the fossil record is characterized by periods of relative stability alternating with short bursts of evolutionary divergence. Often the flowering of new species follows upon the sudden extinction of preexisting organisms. This stop-and-go progression of life is called "punctuated equilibrium."

In this view of biological evolution, mass extinctions play a critical role. If cosmic catastrophes repeatedly disrupted our planet and precipitated paroxysms of dying, they probably also made possible the development of most new species. It is almost as if the course of evolution were dependent on an occasional kick or shake from the outside. Once the Cretaceous event was identified with a cosmic impact, it was natural to inquire whether iridium concentrations or other evidence could be found that might link other extinctions with impacts.

So far no major iridium anomalies have been found in association with the four older great extinctions discussed above, although uncertainties remain. None of these discontinuities has a well-defined boundary layer like that at the end of the Cretaceous, and rocks this old are more difficult to locate and work with than the relatively young sediments from 65 million years ago. These extinctions could instead have been caused by impacting objects that were depleted in iridium or by higher velocity impacts that resulted in ejection of the iridium-rich projectile material back out into space.

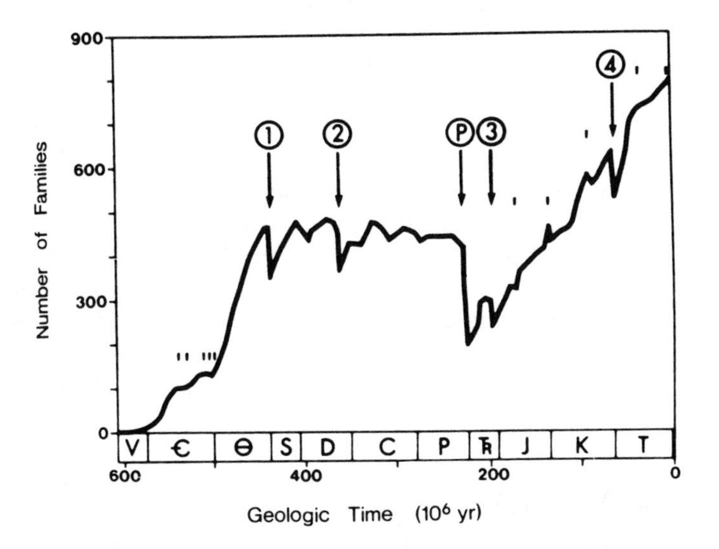

The mass extinction at the end of the Cretaceous (No. 4) is the latest of several mass extinctions that have been recorded since skeletons of marine animals first appeared in the geological record 600 million years ago. This diagram shows at least five sudden declines in numbers of families of marine creatures. (J. Sepkoski)

Scientists have had partial success in finding iridium associated with two smaller mass extinctions that took place since the Cretaceous, both of which are classed as second-rank, which means that less than 50% of marine species perished. The most recent iridium-enhanced boundary was deposited, in at least some places, just 12 million years ago, at the end of the middle Miocene age. The second enhancement recorded at some sites is close in time to the mass extinction at the Eocene–Oligocene boundary, 38 million years ago, where there are layers of glassy spherules ejected from impact events. But it seems that the iridium was deposited several million years *after* the Eocene extinction, which casts doubt either on Eocene dating measurements or on the association of iridium with the extinction.

One intriguing aspect of these three well-documented recent mass extinctions is their spacing. They took place 12, 38, and 65 million years ago, at intervals of about 26 million years. Is this just a coincidence, or might we be seeing evidence of regularly spaced, periodic impacts and their associated mass extinctions? No one had previously imagined that impacts on the Earth or Moon might have followed such a regular pattern, but here is a possible indication of such an effect. If impacts and mass extinctions are actually regular and periodic rather than random, our view of these events, and of the cratering history of the solar system, would be greatly modified.

* * *

Let us look at the statistics of mass extinctions to see if there is additional evidence that they are evenly spaced in time. The standard intervals into which we divide geological time were established more than a century ago, although the assignment of chronological ages to these "periods" and "eras" is more recent. Some (but not all) of the divisions between these time intervals correspond to mass extinctions, and some (but perhaps not all) such mass extinctions could well have been caused by impacts. It is apparent from a cursory examination of the geological time scale that many of these divisions are between 20 and 40 million years in length. Most scientists have assumed, however, that this spacing was an artifact of the naming process itself, representing an interval that was convenient for classification and nothing more. The suggestion that these divisions are not artifacts, and that mass extinctions are truly spaced at regular intervals, was first made in the early 1980s by David Raup of the University of Chicago, a past president of the Paleontological Society and a member of the National Academy of Sciences, and by his younger Chicago colleague Jack Sepkoski, who had completed his doctoral work under Steven Jay Gould at Harvard.

Both Raup and Sepkoski are experts in statistics and have earned reputations for their careful analysis of extinction rates in the geological record. Their research utilized a new computerized data base that listed the times of first and last occurrence in the geological record of thousands of individual families of marine organisms. (More recently, Sepkoski has been working on a similar catalog at the genus level, which contains tens of thousands of listings.) In 1983, they began to see a pattern in the recorded times of extinction of the families in their data base. The mass extinctions seemed to come at intervals of 25 to 30 million years. They were aware, however, of the pitfalls in verifying such regularity, since even in a random set of data, patterns often seem to emerge, just by the luck of the draw. They needed a sophisticated computer analysis to determine if the apparent regularity represented statistically sound evidence of periodicity that was capable of standing up to informed criticism by other scientists.

By 1984, Raup and Sepkoski were sufficiently confident of their conclusions to publish a paper in the *Proceedings of the National Academy of Sciences* entitled "Periodicity of Extinctions in the Geologic Past." Their paper asserted that 12 mass extinctions (8 substantial events and 4 smaller ones of marginal significance) have taken place in the past 250 million years, and that these events repeat in a 26-million-year cycle. The illustration shows how well the 8 major extinctions match the 26-year interval. Two of the expected events, number 5 (120 million years ago) and number 7 (170 million years ago), seem to be missing, but the others are quite evenly spaced. The most recent 3 extinctions are those we mentioned before as possibly associated with iridium enhancements, and thus likely involved extraterrestrial impacts.

Since the original publication of these results, many scientists have joined the debate about the reality of the reported periodicity. Critics noted that the mass extinctions were based on the time history of just 567 families of fossil organisms. By 1988, Sepkoski had extended the analysis to a larger data base of 11,000 genera, which demonstrated an even more impressive periodicity, but only since the missing event 170 million years ago. Another criticism concerns uncertainties in the exact chronological dates of the extinctions. The

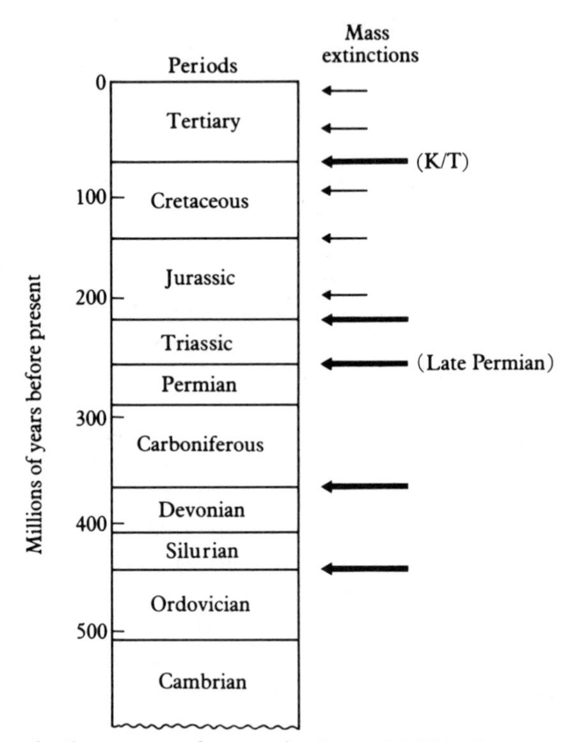

Have mass extinctions occurred at regular intervals? This diagram summarizes the major and minor mass extinctions that have been established in the geological record of the past 500 million years of Earth history. (From *The Planetary System* by D. Morrison and T. Owen, Addison-Wesley Publishing Company, 1988, based on data from D. Raup in *The Nemesis Affair*, Norton, 1986)

geological time scale is defined by the fossil-bearing layers in marine sediments, but it is not simple to date these well-defined boundaries as having specific ages, which must be measured from the residual radioactivity of volcanic ash or other igneous rocks buried in the largely sedimentary geological column. The uncertainty in timing the end of the Cretaceous is probably only a million years either way, but the older mass extinctions could be misdated by as much as 10 million years, and there are also uncertainties about the more recent Eocene. If the dates of the extinctions are uncertain, it is difficult to establish the reality of their periodicity. Raup and Sepkoski believe, however, that errors in the dates of extinctions are unlikely to lead to a false appearance of periodicity. Large errors might mask a truly periodic effect, but it is very difficult to see how random errors could introduce the illusion of regular spacing if none were really present.

In 1983, as word of the Raup and Sepkoski analysis circulated within the scientific community, several groups of geologists and astronomers were inspired to examine the growing data on the ages of terrestrial impact craters for possible indications of periodicity. By this time about a hundred astroblemes had been identified on our planet. Is there direct evidence, from the large craters themselves, of a regular 26-million-year spacing? A first analysis, carried out by Raup and Gene Shoemaker, turned up no indication of regular spacing from craters with known ages. But in April 1984 two groups published papers reporting positive results. Geologists at Berkeley (including Walter Alvarez) found evidence of a 28-million-year periodicity for terrestrial impact craters, while two NASA astronomers found a period of 31 million years, working from a somewhat different selection of crater age data. Although the differences between these two derived periods and the 26-million-year period found by Raup and Sepkoski may seem significant, they probably are not, given the uncertainties in the measured ages of the craters.

The new results seemed to confirm both the association between impacts and extinctions and the periodicity of these events, but, as usual in science, objections were raised. Only about 20 ancient craters have been dated, and their ages are even more uncertain than those of the boundaries between geological periods. The estimated uncertainties in many of the crater ages amount to 10 million years or more. With such large uncertainties in the basic data, the statistical analysis is not very convincing. It thus remains an open question whether the evidence from terrestrial impact craters supports the suggestion of periodicity or not.

One of the most surprising new aspects to this puzzle is the possibility that large impacts are multiple. There are at least two iridium-rich layers dating from 35 million years ago, near the Eocene–Oligocene boundary, separated by intervals of at least tens of thousands of years. Also, about seven glassy ejecta layers have been recognized within a 2-million-year period around the end of the Eocene. It would seem to be a remarkable coincidence for numerous separate large impacts to have occurred within so short a span of geological time. Perhaps this is no coincidence, and the 25- to 30-million-year periodicity actually involves many projectiles. Maybe the Earth was subjected to periodic comet or asteroid "showers," rather than single impacts.

* * *

We have now described the components of a remarkable scenario for the history of our planet and the evolution of its life-forms. Let us briefly recapitulate. It is generally accepted that biological evolution follows the pattern of punctuated equilibrium, with mass extinctions playing a major role in initiating the flowering of new families, genera, and species. In the case of the Cretaceous–Tertiary mass extinction, the killing resulted from the impact of one or more large extraterrestrial projectiles, and a convincing circumstantial case is being made that similar impacts are implicated in many (perhaps most) other mass extinctions. Although not everyone is convinced, Raup and Sepkoski have argued that the mass extinctions, and hence the impacts, recur regularly with a period of 25 to 30 million years. Finally, there may be several large impacts associated with each mass extinction, giving rise to the concept of periodic comet or asteroid showers. Here is a cosmic connection indeed!

If the Earth is bombarded periodically by comets, asteroids, or other cosmic projectiles, there must be some astronomical cause or driving force. Within weeks of the announcement by Raup and Sepkoski of their evidence for periodic impacts, the search began to identify an astronomical driving force that might operate on a 25- to 30-million-year time scale.

Now 25 million years is a very long time, even in astronomy. The solar sunspot cycle is a mere 11 years; the planet Pluto requires 248 years to circle the Sun; and even the precession of the Earth's axis (the phenomenon that causes the apparent location of the north celestial pole to move through the stars) has a cycle of only about 26,000 years, too short for our purposes by a factor of 1000. Nothing we know about the solar system itself oscillates on a time scale measured in millions of years. Only if we look outward to the great Milky Way Galaxy of stars, of which our Sun is a part, can we find time scales of the requisite magnitude; for example, the "galactic year," the time required for the Sun and solar system to make one circuit around the galaxy, is about 250 million years—ten times greater than the impact cycle we would like to understand.

One of the most straightforward ways to explain impacts spaced at intervals of 25 to 30 million years is to hypothesize the presence of an undiscovered object in orbit around the Sun with this orbital period. No one is suggesting that this mysterious object itself causes the impacts, but it might serve as the trigger or catalyst. Recall that the comets we see represent leakages from the giant comet reservoir, the Oort cloud. This cloud of about a billion billion comets is made up of objects that orbit the Sun with periods of the order of 100,000 years and typical distances of about a light year—a third of the way to the nearest star. According to the model first suggested by Jan Oort in 1950, passing stars occasionally, and randomly, perturb a few of the comets in the cloud so that their orbits bring them close to the Sun, where we can see them. But suppose that in addition to this random source, an interloper brushed the edge of the comet cloud at regular intervals and dislodged thousands of comets, causing them to fall into the inner solar system. Such comet showers, persisting for hundreds of thousands of years, are just what is needed to generate enhanced impact rates periodically.

Several astronomers proposed the existence of a small undiscovered companion star to the Sun, in an orbit that might lead to partial disruption of the Oort comet cloud every 25 to 30 million years. This hypothetical object is often called "Nemesis," as explained in one of the two independent scientific papers (both published early in 1984) in which the idea was first suggested: "If and when the companion is found, we suggest that it be named Nemesis, after the Greek goddess who relentlessly persecutes the excessively rich, proud, and powerful." As an alternative name, Steven Jay Gould has proposed Shiva or Siva, for the Hindu god who periodically destroys and recreates the world: "Unlike Nemesis, Siva does not attack specific targets for cause or for punishment. Instead, his placid face records the absolute tranquillity and serenity of a neutral process, directed toward no one." Others have called this conjectural agent of destruction simply the Death Star.

The only aspects of Nemesis/Siva, the Death Star, that can be securely specified are its orbital period—25 to 30 million years—and the fact that it must last have brushed the comet cloud about 12 million years ago, at the time of the most recent impact-induced mass extinction. From the inferred orbital period it is straightforward to calculate that the average distance of the Death Star from the Sun must be about 2 light years, or roughly half the distance to the nearest known star, Alpha Centauri. Since the last comet shower took place 12 million years ago, the Death Star should now be far from the Sun, receding slowly into space. If it exists, it should reverse its motion and begin returning in another few million years, passing through the comet cloud again about 15-odd million years in the future.

The Death Star need not be very large to provide enough of a gravitational nudge to stir up the comet cloud when it brushes past. One or two percent of the mass of the Sun is probably sufficient. Thus the Death Star could be either a small star of low mass (called a "red dwarf") or even a substellar object. Calculations have shown that the minimum possible mass for a star—defined as an object that emits energy produced by thermonuclear reactions in its core—is about 1/12th the mass of the Sun. A smaller object never becomes hot enough to kindle its nuclear fires. Such substellar objects, which glow feebly from the residual heat of their formation, are called "infrared dwarfs." They are very difficult to detect and may be quite rare in nature, but astronomers are confident that such infrared dwarfs exist.

Our hypothetical Death Star thus could be either a faint red dwarf or a still fainter infrared dwarf. It is possible, though unlikely, that such a dim object might remain undiscovered at a distance of 2 light years. In particular, the infrared survey satellite IRAS, which scanned the entire sky in 1983, should have detected such an object if it were present. The data records from IRAS have been searched for such a detection, so far without success.

The other difficulty with the Death Star hypothesis is that an orbit that carries such a companion 2 light years from the Sun is not very stable. When near its greatest distance, the solar companion is so weakly held by the Sun's gravity that it could easily be dislodged by the pulls of other nearby stars. The coupling between the Sun and its faint companion might survive for a few hundred million years, but not for billions of years. And even if some gravitational bond survived, it seems unlikely that the orbital period would remain stable. Ironically, the apparently very regular periodicity of the mass extinctions may be too precise to lend itself to the Death Star explanation, and thus becomes an argument against the hypothesis.

If there is no Death Star, what other astronomical driving force might be invoked to perturb distant comets at regular intervals? There is one possibility. In addition to its orbital path around the galaxy, our Sun bobs slowly up and down, perpendicular to the plane of the galaxy, as do other stars. Like an excruciatingly slow pendulum, the Sun rises above the galactic plane and then is drawn back down by gravity, passing though the central plane and emerging on the other side, where again its motion is reversed. This oscillatory motion is too slow to be observed, but scientists estimate its period, from the measured distribution of matter in the galactic plane, to be about 60 to 70 million years. Thus the Sun and solar system pass through the galactic plane about once every 30 to 35 million years, which is close to our sought-after cycle of 25 to 30 million years.

The time scale for the Sun's oscillation through the galactic plane is about right, but how does this motion lead to comet showers? One possibility is that there is a dense region of interstellar debris concentrated in the galactic plane, and it is these external projectiles rather than comets from our own comet cloud that generate the impacts. Alternatively, there may be masses in the galactic plane that can dislodge comets from the Oort cloud, in just the way suggested for the Death Star. For instance, the solar system sometimes passes through or near what astronomers call a "giant molecular cloud," which is a region of relatively dense gas and dust about 100 light-years across. There are many such molecular clouds in the galactic plane, and the Sun has a good chance of encountering one of them when it makes its periodic incursions through the plane.

The similarity of the periods for the galactic oscillation of the Sun and the extinctions is encouraging, even though we don't know exactly how the oscillations would produce comet showers. But there is a major stumbling block. The Sun just passed through the galactic plane within the past couple of million years, and is still close to it. In contrast, the periodicity of the extinctions implies that the Sun should have passed through the plane about 12 million years ago and now be near the farthest point above the plane. The two phenomena—mass extinctions and galactic oscillations—have the same period but are "out of phase." This is a serious objection to the hypothesis; in the

The Great Galaxy in Andromeda is a twin of our own Milky Way Galaxy. Some scientists suggest that motions of the Sun in our galaxy could generate periodic mass extinctions. (Palomar Observatory photo)

opinion of many scientists, the discrepancy is great enough to reject the galactic hypothesis outright.

We are left with a great mystery. There is evidence, circumstantial and incomplete but still credible, that much of the history of our planet and of the evolution of life is dependent on an astronomical source that generates impacts and mass extinctions on about a 30-million-year time scale. Yet we have not identified this cosmic driving force. The search continues, even as scientists seek to verify the periodic nature of mass extinctions and to understand the linkage between them and extraterrestrial impacts.

* * *

One intriguing aspect of the whole debate about dinosaur deaths and comet showers is that it has brought into the same arena a great diversity of scientists. It is difficult to think of any other question of modern science that would involve cratering theorists arguing with paleontologists, evolutionists debating with astrophysicists, atmospheric scientists comparing notes with cosmochemists, and so on. Usually, scientists work in their own narrow specialty, occasionally discussing matters of joint interest with an expert in a related field.

In October 1981, soon after the publication of the Alvarez hypothesis, an unusual meeting was held at the ski resort in Snowbird, Utah, that brought just such a diverse group of scientists together. There have been subsequent large meetings before the American Association for the Advancement of Science, a second Snowbird conference, and small workshops on aspects of the Alvarez/comet-shower hypothesis. But no later meeting could quite capture the wonderment at Snowbird as one faction sized up another as though they were aliens from another planet. To planetary scientists like Gene Shoemaker and George Wetherill, it is as obvious that the Earth is occasionally struck by asteroids and comets as it is that the Sun rises in the East. But the idea seemed outlandish to many of the geological participants; Dewey McLean, from Virginia, argued that the extinctions were clearly associated with volcanism in India. Geochemist Jan Smit, from the Netherlands, reported evidence that some extinctions occurred in less than 50 years (perhaps *much* less) while another speaker argued that they occurred over millions of years; indeed, according to the extreme views of Norman Newell, mass extinctions can hardly be considered catastrophic and may be mostly illusions.

Interspersed between talks about fossilized pollens and plankton were reports about explosion experiments, the nature of comets, and dinosaur bones. Chemist John Lewis described the poisoning of the Earth's atmosphere that might result from a huge meteor. All in all, it was an eclectic meeting of scientific opposites. Over the years since the first Snowbird conference, the various researchers have gotten to know each other better. But many still find it difficult to appreciate the work of their cross-disciplinary colleagues, so debates still rage in the literature.

Already, by the end of the Snowbird meeting, some participants with wider horizons were beginning to see the epochal nature of the joining of paleontology and astronomy mandated by the Alvarezes. Conference summarizer Lee Silver, from the California Institute of Technology, saw the meeting as the converging study of two evolutionary processes—the evolution of astronomical bodies and the evolution of biological systems—mediated through the special environment that is our abode, the planet Earth. Through these studies of how biological systems of the past have responded to extraterrestrial influences on our Earth, scientists came to realize that there are lessons for us in how the increasingly profound influences of civilization on our planet may affect our own future.

CHAPTER 8

Nuclear Winter

The Earth's ecology is a fragile thing. Because of the interdependence of the atmosphere, oceans, climate, and life, the explosive impact of a single comet or asteroid can have devastating ecological consequences, even though the mass of the projectile is less than one billionth of the total mass of the Earth. Thus, an impact that would be of little consequence on the Moon or Mars can disrupt conditions on the Earth in drastic ways. We live in a hair-trigger world, highly vulnerable to injury.

Increased appreciation of this fragility is one product of the Alvarez ideas about the impact that ended the Cretaceous period. Another result has been the recognition that similar ecological consequences could be triggered by global nuclear war. While scientists and policy-makers alike have understood for decades the specific effects of nuclear explosions—the radius of destruction, the number of buildings shattered and burned, the direct casualties and those induced later by radioactive fallout—they had not considered the cumulative effects on the environment of a widespread nuclear attack. Because the atmosphere is so sensitive to perturbation, the totality of these effects would be very much greater than simply adding up the destruction and casualties from each bomb individually.

The concept that global environmental damage could result from a nuclear war has been termed "nuclear winter," since the primary effect would be cooling of the surface due to absorption of sunlight by atmospheric smoke. This term was coined in 1983 by Richard Turco, an atmospheric scientist from a small research organization called R&D Associates in Marina Del Rey, California. Turco combined forces with three planetary theorists at the NASA Ames Research Center south of San Francisco: Brian Toon, Tom Ackerman, and Jim Pollack. The Ames group had experience in computing the effects of perturbations on the atmospheres of the Earth and other planets using computer models that variously simulated the dust cloud of the Cretaceous asteroidal impact, the major annual dust storms on Mars, and the effects on Earth's climate of the 1982 eruption of the Central American volcano El Chichon. As a result of such previous research, this team had the expertise and computational tools to simulate the consequences of different scenarios for injection of dust and smoke into the terrestrial atmosphere.

A fifth member of the nuclear winter team was Carl Sagan of Cornell University, by this time probably the best-known scientist in the United States. Sagan was one of the first of the new breed of planetary scientists when he began theoretical studies of the atmosphere of Venus in the 1960s, but now his fame as a science popularizer has eclipsed his role as an active researcher. Sagan had served as thesis advisor to both Toon and Pollack and had long been interested in comparative studies of the atmospheres of Earth, Mars, and Venus. He wrote an article on the policy implications of nuclear winter that was published in *Foreign Affairs* at the same time as the scientific results came out in *Science*. The *Science* paper appeared at the end of December 1983, signed by Turco, Toon, Ackerman, Pollack, and Sagan: a group of authors with the memorable acronym TTAPS.

The TTAPS paper, like the Alvarez article three years earlier in the same journal, was one of the most controversial and influential scientific publications of the 1980s. It initiated a widespread discussion of the global consequences of nuclear war and stimulated a variety of subsequent atmospheric models, including studies carried out by the Academies of Science of both the United States and the U.S.S.R. Spurred by the public clamor, the Department of Defense authorized the expenditure of $50 million in follow-up studies, although at the same time other major American political figures were asserting that the concept of nuclear winter would prompt no change in United States nuclear policies. There is an interesting difference, however, between the TTAPS and Alvarez papers. By their discovery of enhanced iridium in the Gubbio clays, the Alvarezes and their coworkers introduced critical new data that forced us to modify our view of mass extinctions. In contrast, the TTAPS results involved no new data. They represented purely theoretical findings, stimulated by a new way of asking basic questions about the consequences of nuclear war. No one could question the reality of the iridium measured at Gubbio, but the credibility of nuclear winter depended critically on the accuracy of complex atmospheric models operating on the frontiers of modern computing technology.

Much of the subsequent debate concerning nuclear winter has centered around the differences among computer models for the Earth's atmosphere. Understanding the dynamics of the atmosphere is essential for any long-term predictions of weather, and the economic (and military) advantages of being able to anticipate the weather 30 or 60 days in advance would be profound. Each advance in computers over the past two decades has inspired efforts to increase the sophistication of atmospheric models. In some respects these models have demonstrated great capability, for example, in their ability to simulate the large-scale weather patterns on Venus and Mars as well as on the Earth. However, we still do not have dependable 30-day or 60-day forecasts, and probably never will. The inherent unpredictability of a chaotic system as complex as the Earth's atmosphere defies accurate analyses. Yet these same programs must be used to model the effects of injection into the atmosphere of dust and smoke from a nuclear war, so uncertainties remain about the calculated consequences. Short of starting a nuclear war, the models cannot be tested experimentally, so some doubt must always persist.

Granting these uncertainties, however, all computer models have confirmed the reality of the nuclear winter phenomenon. If many cities burn simultaneously, the load of smoke injected into the atmosphere is sufficient to shroud much of our planet in darkness, lowering the surface temperature and altering the patterns of atmospheric circulation. The models differ on important details, however, such as the degree of cooling and its duration. Also in dispute is the coupling between the northern and southern hemispheres. If (as seems likely) a nuclear war is confined to the northern hemisphere, it is unclear to what extent the dust cloud will spread over the southern half of the planet as well. All of this depends also, of course, on the magnitude of the nuclear conflict itself.

* * *

The atomic bombs that devastated Hiroshima and Nagasaki in 1945 had energy yields of about 20 kilotons, that is, the equivalent of 20,000 tons of TNT. In contrast, the typical strategic hydrogen bomb of today has a force of a megaton (a million tons of TNT), and the United States and the U.S.S.R. each have thousands of delivery systems (land- and sea-based ballistic missiles, cruise missiles, and bombers), many of which can launch multiple warheads. In addition, each side has thousands of smaller "tactical" nuclear weapons, mounted on short-range missiles, fighter-bombers, and even in artillery shells and antisubmarine depth charges. According to the American National Academy of Sciences, "United States and Soviet nuclear forces reportedly now include about 50,000 nuclear weapons, with a total yield of some 13,000 megatons. About 25,000 of these nuclear weapons, with a yield of about 12,000 megatons, are on systems with strategic or major theater missions." It is assumed that the majority of these weapons are targeted on the approximately 1000 underground missile silos in the United States or the 1400 silos in the U.S.S.R. Additional thousands of these weapons are aimed at military targets such as command and communications centers, airfields, naval bases, and transportation nodes. The philosophy of attacking these military targets is called "counterforce." This still leaves several thousand bombs for "countervalue" attacks—those directed against targets of economic significance such as factories, refineries, seaports, dams, and electric power plants. It is not known if today's targets include civilian population centers *per se*, but obviously the counterforce and countervalue targets would include most urban areas in both nations. Yet another frightful possibility is targeting of nuclear power plants, which would increase the total radioactive fallout from a nuclear attack severalfold over that produced by the hydrogen bombs themselves.

The meaning of such statistics is difficult to absorb. There are only about 500 urban or metropolitan areas with a population of 100,000 or more in the United States and the U.S.S.R. Even if only 10% of the nuclear arsenals of each side were targeted at such areas, there are 5 bombs per city, each of them several times larger than the Hiroshima or Nagasaki atom bomb. Another way of looking at this awesome level of overkill is to note that 13,000 megatons of yield corresponds to 26 tons of high explosive for every single Russian and American alive today. By comparison, the United States–Soviet agreement in the INF treaty of 1987 to eliminate 4% of their strategic forces pales into insignificance, except for the symbolic hope it provides.

In order to study the global effects of a nuclear war, scientists must make some assumptions about how this arsenal of weapons might be used. For the purpose of comparing models, the Academies of Science of the United States and the Soviet Union have agreed on a baseline that assumes that approximately one half of these weapons might be used in an all-out war involving not only the two superpowers, but other NATO and Warsaw-Pact countries as well. Additional assumptions must be made about counterforce vs. countervalue strategic attacks and about the use of tactical or battlefield weapons. Because of their small yield, the tactical weapons contribute little to the global situation. It is assumed that most of the counterforce weapons would be exploded at ground level ("groundbursts") in order to damage well-protected missile silos and other underground installations, while some counterforce and most countervalue weapons would be exploded about a mile above the surface ("airbursts") to inflict maximum radiation and blast damage on economic and urban targets. In the baseline war, more than 10,000 individual nuclear explosions would take place within a day or two, almost all of them in the northern hemisphere. Even without specifying individual targets, we can be pretty sure that most if not all urban areas in the United States, the U.S.S.R., Europe, and Canada would be destroyed. With this baseline scenario, we can undertake to estimate the consequences.

Each groundburst explosion would generate several tens of thousands of tons of fine dust particles that would be lofted into the stratosphere by the fireball. In addition, such groundbursts would probably start extensive wildfires that would consume large parts of the grass and forest land that typically constitutes the locales for military bases in the United States and the Soviet Union. But far greater damage to the environment would result from the smaller number of warheads targeted toward industrial, communications, and population centers.

An airburst of a 1-megaton bomb over an urban area results in total destruction to a radius of several miles, with partial collapse of buildings and the setting of extensive fires out to about 10 miles from "ground zero." As we know from Hiroshima and Nagasaki, and even from the "conventional" World War II firebomb raids on Tokyo, Hamburg, and Dresden, the result is likely to be an intense firestorm, with hurricane-force winds creating heat tornadoes, which feed the flames and coalesce into towering convective plumes of smoke. Almost all flammable materials, including a wide variety of toxic chemicals, would be consumed in such a firestorm. A large-scale urban firestorm would inject several million tons of smoke and soot into the upper atmosphere.

One of the discoveries that led to the concept of nuclear winter concerns differences between the optical properties of soot and ordinary dust. Most of our experience in calculating the atmospheric and climatic effects of particulates is associated with dust produced in volcanic explosions or in a meteoritic impact. But it happens that soot particles, because of their different composition, are much more effective than dust in absorbing sunlight. In fact, a given mass of fine soot or smoke transmits only about 1% as much sunlight as an equivalent mass of dust. The most serious environmental effects of nuclear war are derived from the stratospheric soot produced from burning cities and wildfires.

In the baseline scenario for nuclear war, the upper atmosphere over the temperate northern hemisphere receives an injection of 50 to 100 million tons of dust and 200 to 300 million tons of soot, mostly particles less than a micrometer in size (about 1/25,000th of an inch). These clouds, which rise as high as 15 miles above the surface, are totally opaque, turning day into night over much of the devastated land. In the inferno below, radioactive fallout and toxic gases from the burning cities contaminate much of the land and add 100 million human casualties to the more than 200 million already killed by blast and fire. Most of the United States, the Soviet Union, and Europe is a charnel house, incapable of supporting human civilization for generations. But what of the rest of the Earth? That is where the calculations of global atmospheric effects become important.

* * *

Under normal circumstances, most sunlight reaching our planet is absorbed at the surface, where it warms both the ground and the lower atmosphere. There is generally an upward flow of energy, which keeps the bottom 6 to 8 miles of the atmosphere—the troposphere—in a state of continual convective circulation. The troposphere is where water clouds condense and thunderstorms form. Above it, the much more placid stratosphere extends upward, clear and cold, for about 20 miles. At night, the direct heating of the surface ceases, but heat continues to flow upward from the land and oceans, maintaining the basic daytime structure of the atmosphere.

Stratospheric clouds of dust or soot would greatly alter this picture. If the soot is thick enough to absorb most of the incident sunlight, the stratosphere will be heated, and the ground below will cool. For a large, dense cloud of soot, the TTAPS calculations show that after a period of a few weeks, the atmosphere 15 miles above the surface will warm by about 100°F while the surface cools by 30 to 50°F. The normal upward energy flow will cease, and the troposphere as we know it will nearly disappear. The stratosphere will descend to the surface, while the hot black clouds will tend to rise and expand, fundamentally altering the dynamics of the atmosphere. Because these conditions are so unlike those normally encountered on Earth, the computer models must be carefully constructed if they are to simulate these new conditions realistically.

The soot cloud from a single urban firestorm would be expected to dissipate within a few days, due to both fallout (and washout by rain) and lateral mixing with clean air. The effects would therefore be restricted to a small area and would quickly pass. But if the soot from a hundred such urban firestorms is combined, the clouds will coalesce, spreading their black pall over millions of square miles. In the aftermath of a full-scale nuclear war, we would expect these clouds to blanket most of the northern hemisphere between the latitudes of 30° and 70°. While the larger particles of airborne soot and dust will settle or be washed out by rain within a few days, the more numerous small particles (less than 1/25,000th of an inch in diameter) in the stratosphere will remain for a few months, just as we have seen for the end-Cretaceous dust cloud. Since it takes only a couple of weeks for stratospheric circulation to carry a temperate-latitude cloud entirely around the Earth, all of these clouds will soon join together. Much of the continental surface will cool by 30 to 50°F, although nearer the coasts the moderating effects of the oceans should limit the cooling. When the cloud gradually clears, the temperatures should return to more normal values, probably within 3 to 4 months.

We can well imagine the consequences for the Earth of such a globe-shielding black cloud. Initially, the bomb-shattered land is immersed in a vast gloomy darkness. Within a week the inland temperatures drop dramatically. Obviously such a cooling will be hard on surviving life-forms that are not protected, although within the war zone most large animals may already be dead from blast, fire, and radiation. However, the effect of the nuclear winter on plant life depends critically on the season in which the nuclear war is fought. If the nuclear winter is simply superimposed on a real winter, the effects on temperate-latitude ecology are relatively small. Most plants and many animals spend the coldest months in a dormant state and are thus equipped to survive even an extremely severe winter. But a war fought in spring or summer would have quite a different effect.

Imagine a sudden drop of 40°F just when life is in flower during the long, warm days of spring and summer. Within a few days freezing rain and hail would begin to fall, and another week would bring deep blankets of snow. Unprepared for such a change, most plants would perish, and with them the animal life that depends on them for food. A drop of even as little as 20°F during the growing season would devastate the forests and grasslands and completely eliminate an entire season of crops. (Because they have been bred for high yield under controlled conditions, cultivated crops are much more easily damaged than natural ecosystems.)

We can see the importance of understanding how the stratospheric soot could extend from temperate regions to the tropics and how much would cross the equator into the southern hemisphere. Very few species endemic to the tropics could survive such a cold snap. And of course, the southern hemisphere has opposite seasons from the northern, so that a winter war in the north might result in summer destruction in the south.

Another question, the answer to which depends on which computer model you believe, is the degree to which noncombatant nations would be damaged by large-scale nuclear war between the United States and the Soviet Union. An exchange involving thousands of weapons would surely amount to mutual suicide for the superpowers. But what of the rest of the world? The calculations so far suggest that climatic effects would also disrupt the economies of at least Canada, Mexico, China, Japan, and the nations of Europe, with death by starvation and disease for perhaps a billion of their inhabitants. If, in addition, the cloud expanded south of the equator before dissipating, the toll could extend to most of the Earth, creating a truly global catastrophe. We have long known that there can be no victors in a full-scale nuclear war; now it seems that there can be no noncombatants either.

The atmospheric calculations, and the controversies surrounding nuclear winter, continue. Heated and sometimes unseemly arguments take place between those who conclude the maximum surface cooling would be 30°F for 2 months or 50°F for 4 months. This unusual level of passion in a scientific debate reflects the high stakes, which are no less than the future of civilization itself. But this debate

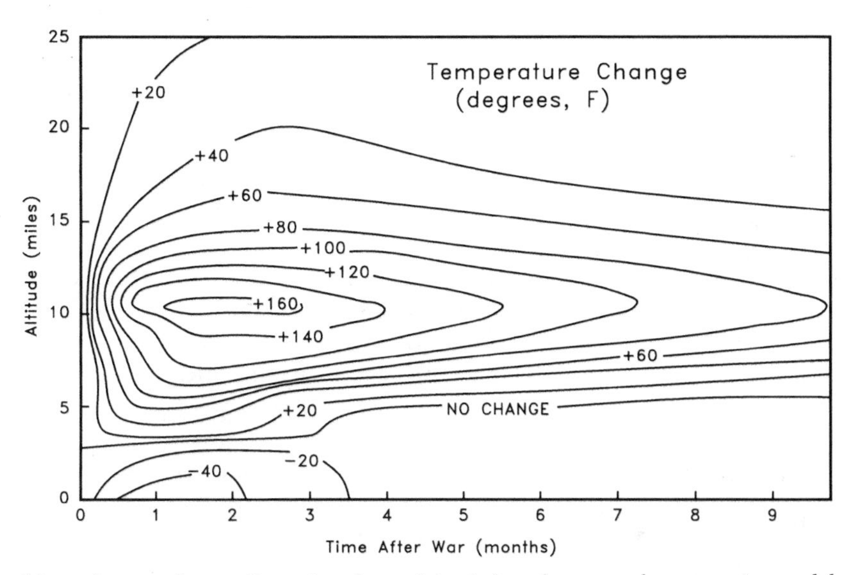

Most of our understanding of nuclear winter is based on complex computer models that simulate the climatic effects of the deposition of massive quantities of soot in the stratosphere. Results plotted here show heating of the upper atmosphere, and cooling of the lower atmosphere, during the months following a nuclear war. (Based on data from B. Toon)

over details should not obscure the basic fact that the minimum short-term result of a nuclear war would be a climatic catastrophe with possible loss of crops for a full year, adding the horrors of mass starvation and disease to the previously catalogued consequences of the nuclear weapons themselves.

Nuclear war is a man-made disaster. Why do we include it in a book on cosmic catastrophes? First, these recent advances in understanding the horrible consequences for our planet have come from research by planetary scientists familiar with studying the other planets. Perhaps it takes a cosmic perspective to appreciate, and deal with, the global consequences of such a man-made Armageddon. But in a larger sense, the evolution of life and development of civilization on our world is itself a cosmic process. When living things evolve to the point where their activities can transform the very climate of their planet, they emerge into cosmic significance. Soon human civilization may develop the capability to "terraform" other worlds in potentially beneficial ways (though a new breed of planetary environmentalists might object). Although there has been thinking about possibly terraforming Mars into a habitable world, it seems that our present level of potential cosmic impact is limited to the destructive. Since the nuclear age began, we have become cosmically significant, but so far only for our potential to create our own cosmic catastrophe.

* * *

Let us conclude by comparing nuclear winter with the other catastrophes we have discussed, particularly with the impact that ended the Cretaceous period. The energy of that 65-million-year-old impact was estimated at 100 million megatons, and the mass of material excavated from the crater at 100 trillion tons. In comparison, the baseline scenario for nuclear war discussed above involves the explosion of only 11,000 megatons and the production of less than a billion tons of dust and soot. The two events seem so different in scale as to call into question the similarity of their climatic effects.

Of course, the explosive energy in itself is not that important. In both examples, the global damage to our planet comes not from the direct effects of the explosions but indirectly, through disruption of Earth's climate and ecology. Further, the blast damage to the Earth's surface is proportionately much greater in a nuclear war because it is distributed over many selected targets, not dissipated in a single large explosion. Our main focus in this comparison, however, should be on the material injected into the atmosphere.

The baseline nuclear war scenario results in about 200 million tons of fine soot in the stratosphere, which can absorb as much sunlight as about 20 billion tons of similar-size dust particles. For comparison, the impact 65 million years ago injected 100 trillion tons into a global dust cloud, as we can reconstruct from the measured thickness of the boundary clay layer. Thus the nuclear winter black cloud is less effective at absorbing sunlight than the cloud that killed the dinosaurs by a factor of several thousand. The end-Cretaceous impact was a close thing, very nearly wiping out terrestrial life entirely. In comparison, the devastation of a nuclear winter is much less significant for the planet as a whole, for while it might result in the death of a majority of the individual land animals and plants alive at the time, few if any species extinctions would occur. On the geological time scale of hundreds of thousands or millions of years, the overall biology of the Earth would be little affected by nuclear winter. This fact, however, would provide little comfort for the billions of humans who might perish from cold and hunger as the result of this minor perturbation in our climate.

There is an event taking place today, however, that will leave a prominent long-term record in the geological column, whatever the ultimate fate of humanity. We are, at this moment, in the midst of a mass extinction resulting directly from human rapacity. Many examples are familiar to us: extinct or nearly extinct species such as the passenger pigeon or the sperm whale, and the many other animals on "endangered species" lists around the world. Less apparent but increasingly important is the indirect loss of species due to the destruction of habitat, especially in the tropical rain forests where the majority of the individual species live, many of them still undiscovered. We may not weep for the loss of an obscure Amazon insect as we do for the Siberian snow leopard or the Chinese giant panda, but the loss is no less real. It reduces the genetic diversity of life on our planet, limiting the possibilities for future evolution. During the 50 years between 1970 and 2020, one half of the species on Earth will have been wiped out forever, more than a million of these from the Amazon basin alone. We are, indeed, the great destroyer, even without nuclear war.

CHAPTER 9

Violent Volcanoes and Mangled Moons

Onward to Saturn! It was Thursday, March 8, 1979, and the historic Voyager 1 spacecraft was 3 million miles beyond Jupiter and headed out. For the final time in a momentous week of discoveries, scientists gathered before the press corps in the Jet Propulsion Laboratory's Von Karman auditorium. Reporters asked their last questions, the exhausted scientists opined yet again about all their new pictures of Jupiter's unusual moons, and "Encounter Week" at the Jet Propulsion Laboratory was over. Or so it seemed.

One of the Voyager engineers, Linda Morabito, was still hard at work that Thursday, cataloging the pictures. On her screen, a smudge appeared in one of the images of Jupiter's pizza-colored inner moon Io. No matter how much she twiddled at her console, she could not make the smudge vanish. It turned out to be one of the most historic pictures ever taken from space, a portrait of an immense volcano caught in the act of erupting. During the following week, seven more volcanoes were found spouting off in Voyager's pictorial record of Io.

It was a remarkable revelation about the solar system. Consider our own world: If an alien spacecraft were to reconnoiter Earth on some typical day, there would be no volcanoes erupting at all. To be sure, there are places—like Hawaii Volcanoes National Park—where magma often oozes from the ground, and little fire fountains play across the black, glassy lava flows. But imagine an eruption propelling sulfur dioxide 100 miles above the surface. Only on very rare days when a Mt. St. Helens explodes, or an El Chichon erupts, might the aliens discover volcanic plumes lofted into our own stratosphere,

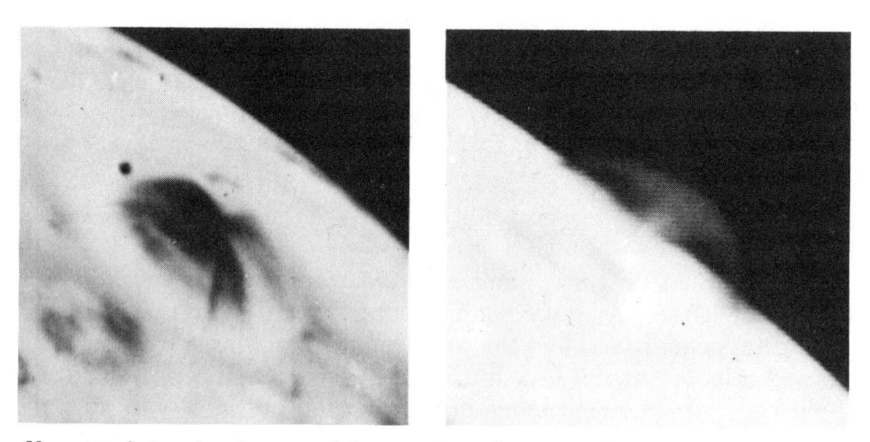

Voyager photos showing one of the erupting volcanoes on Jupiter's moon Io as seen from two different perspectives. (NASA/Jet Propulsion Laboratory)

still shy of the towering Ionian volcanoes. And on Io, such eruptions occur in half a dozen regions at once, all the time!

The discovery contrasted all the more with the growing impression from the first 15 years of planetary exploration that our neighboring worlds were just a collection of crater-scarred corpses. The Moon and Mercury had turned out to be geologically lifeless. While pockmarked Mars begrudgingly revealed signs of volcanism, and even hints of ancient rivers, the Viking landers had just a few years earlier proven Mars to be biologically lifeless.

As the two Voyager spacecraft struck out toward the dark, frozen outer solar system in the late 1970s, most scientists were expecting more of the same: inert, lifeless bodies. Then, in 1979, they were confronted with a small world that was geologically much more alive than even our own planet. Io appeared to be literally turning itself inside out in a turmoil of sulfurous volcanic activity. And that was not all. Io's neighbor in space, Europa—the next moon outward from Io in Jupiter's system—showed a bland surface, lacking even the usually ubiquitous impact craters. Although no volcanoes were visible, something had to be actively freshening the surface of Europa, because evidence of cratering projectiles in near-Jupiter space was all too obvious in the pummeled landscapes on Ganymede and Callisto, two other moons photographed by the Voyagers during their reconnaissance of Jupiter.

Other surprises awaited the Voyager scientists as the twin spacecraft closed in on Saturn in the early 1980s. Scientists felt sure that Saturn's colder, smaller moons would be crater-scarred and inert. To be sure, there were no volcanoes erupting as the Voyagers explored the ringed planet's retinue of moons. But one satellite—the small world Enceladus—showed tracts on its globe that were practically as smooth as the surface of Europa. Evidently Enceladus was no leftover ice-ball from aeons past, either, but had internal geological processes of its own that were rejuvenating and flattening its surface.

In early 1986, just a week before the fateful Challenger disaster back on Earth put the United States space program on hold, Voyager 2 sailed past Uranus, and pictured its small moon Miranda as one of the weirdest places in the solar system. Scientists still don't know quite how to explain Miranda, but it presents the most vivid example yet to counter the early supposition that the small outer-planet satellites would be ancient, cratered, icy spheres, devoid of geological activity or interesting landscapes.

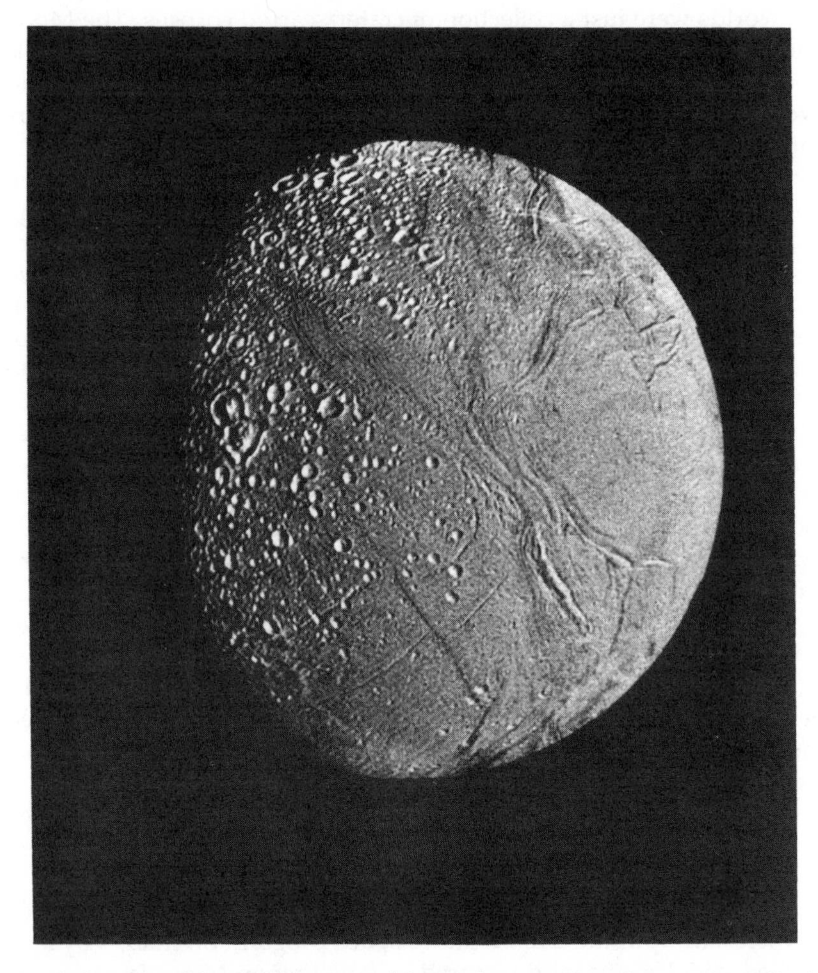

Enceladus, one of the smaller moons of Saturn, has smooth surface terrains indicative of high past levels of internal activity, sufficient to have erased the otherwise ubiquitous impact craters. (NASA/Jet Propulsion Laboratory)

* * *

The Voyager experimenters had not thought very much about the outer-planet satellites before the Voyagers' data began to startle them. The mission was conceived in the early 1970s, as a grand tour of the outer solar system, to take advantage of an unusual planetary alignment that could be utilized to propel the spacecraft to the outermost planets. At that time, it was the gas-giant planets themselves, plus Saturn's fabulous ring system, that were the prime motivators for the missions. The moons were mere specks of light in telescopes, smaller than the smallest planet in our solar system, and hardly worth bothering about.

Until Voyager, simple theoretical notions had helped to keep the wet blanket on these satellites. After all, the things that make our own planet alive—geologically and otherwise—are its comparatively large size and its proximity to the Sun. Also, the Earth has a rich complement of radioactive uranium, thorium, and potassium. Helped by Earth's large size, its radioactive endowment has been more than sufficient to keep the interior warm, providing for the subterranean magmatic activity that fuels terrestrial volcanoes. Indeed, the Earth's internal engine keeps the pattern of geological activity termed plate tectonics rolling along after more than 4.5 billion years. Of course, the Sun keeps the Earth's surface warm and our rivers flowing and eroding, and it maintains a habitat for the biosphere, including ourselves.

All of these conditions should be lacking for a small satellite in the outer solar system. Made largely of ice, such satellites were expected to have small endowments of long-lived radioactive elements, which normally reside in rocky minerals. What little internal heat is generated should leak rapidly away, due to the satellites' small sizes and, thus, their smaller heat-retention capacities. Finally, their surfaces should be frozen solid, with the Sun too distant to warm even dry ice.

Why, then, are several of the outer-planet satellites so geologically alive? There seem to be two answers. First, scientists are beginning to appreciate that some of these moons may be made of especially volatile icy materials that melt at very low temperatures. So while it may be cold on Enceladus for *us*, it is broiling hot for a hunk of ammonia- or methane-rich ice conglomerate. Therefore, on these tiny worlds, normal water-ice may behave as rock does on our world, while the lower-temperature ices behave as water does on Earth. During particularly warm episodes, the rocklike water-ice may melt, creating aqueous volcanism (that is, running water) akin to the silicate magmatic activity and lava flows on our world.

The second reason for churning geological life within some of these bodies is that they are caught in powerful gravitational tugs-of-war between their primary planets and other satellites. Io, for example, orbits mighty Jupiter trapped in step with the resonating orbital periods of some of its sister satellites. While the repetitious gravitational pulls of the other satellites try to keep Io's orbital period fixed, the tides raised on Io's surface, and deep within it, by the enormous gravity of nearby Jupiter are trying to push Io away from Jupiter, for the same reasons that cause the Moon to recede gradually from Earth. The tidal wrenchings generate enormous heat within the small body, which spawns the violent volcanism that shoots sulfurous plumes heavenward. We know this now, but it was not expected as Voyager closed in on Io in 1979.

Nevertheless, before Voyager arrived, Stan Peale, Pat Cassen, and Ray Reynolds had it all figured out. The three California scientists are theoreticians, more comfortable with numbers, equations, and computers than with direct observation. Perhaps their interest in Jupiter's satellites was piqued because Voyager was en route. Whatever their motivations, in late 1978 they were hard at work trying to understand the orbital and rotational dynamics of planetary satellite systems. That is, they wanted to learn how gravitational interactions among Jupiter's satellites affected the satellites' motions. And a few months before Voyager arrived at Jupiter, they realized the implications of the enormous tidal vise-grip in which Io was caught. Hurriedly, they wrote a paper which was printed in *Science* just days before Voyager closed in on Jupiter. In the paper, they predicted that "widespread and recurrent volcanism might occur" on Io and that its effects "may be evident in pictures of Io's surface returned by Voyager." This was a bold prediction, indeed, given the scientific tenor of the times.

Confirmation of Peale, Cassen, and Reynold's prediction was a stunning vindication of theoretical planetary research. The only problem was that, if anything, Io's volcanism exceeded the maximum rate of heat generation (10 trillion watts) they had predicted. Peale and his colleagues were hard-pressed to derive still more heat from Io's tidal flexing. They found it a bit difficult to tally up Io's entire heat budget. Also, there must be some waxing and waning of Io's activity. So the average rate of Ionian volcanism, over the centuries, may be lower and in step with the theory, after all. Ground-based astronomers belatedly discovered that they can measure Io's eruptions from the ground, but they have been monitoring it for only a decade and cannot know if Io is in a temporarily active phase. More surprises may yet be in store for us from Io.

Io's volcanism gives rise to a fantastic array of phenomena associated with Jupiter. Some of the sulfur, potassium, and other products of the eruptions escape into space, where they interact with Jupiter's voluminous magnetosphere—that zone where energetic charged particles are controlled by Jupiter's strong magnetic field, which is generated within the planet and envelops space around it. In fact, Ionian material forms an immense, glowing, doughnut-shaped torus around Jupiter. Interactions between Io, Jupiter, and its space environment control Jupiter's natural radio emissions, which are broadcast throughout the solar system, and which are second only to those of the Sun in power.

It had seemed to the theorists—who calculated tidal heating effects for other planetary satellites and found them very small compared with those of Io—that the surprises would be over. But we have enigmas with other satellites, too, and tidal heating may well play a role in them, as well. All in all, Io, Europa, Enceladus, and Miranda are nearly as complex and interesting from a geological perspective as our own planet Earth. But heat-driven volcanism is not the only force that has influenced the smaller bodies of the outer solar system.

* * *

During the Mariner and Viking explorations of Mars, pictures were snapped of Phobos, the larger of the red planet's two small moons. One side of the oblong body was found to be gouged out by a crater more than half as big as Phobos itself. Radiating away from this crater, dubbed "Stickney," were globe-girdling "grooves"—actually trenchlike valleys—with a form never before seen on any planet. It looked as though Phobos had been struck by a comet or asteroid with nearly enough energy to break the small moon apart. The gaping crater Stickney was the result, with the grooves marking the fractures that almost split Phobos into pieces. Apparently Phobos was unlucky enough to have suffered such a big impact, but luckily escaped being

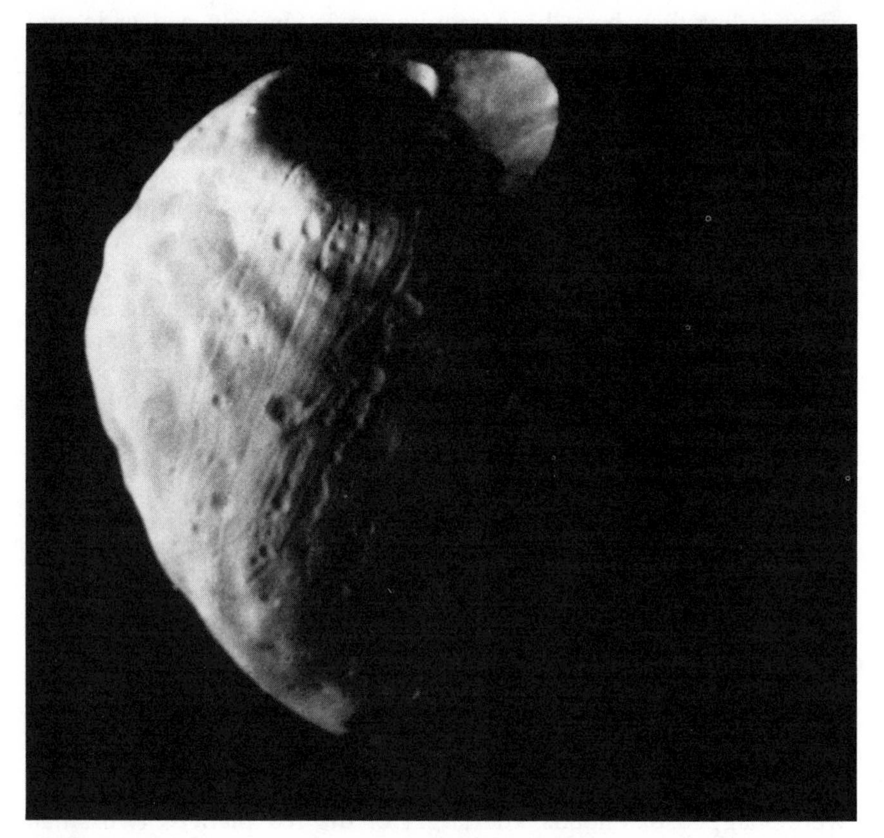

Phobos, one of two moons of Mars, probably looks a great deal like many of the smaller asteroids. The grooves in its surface may have resulted from fracturing by the large impact crater, perhaps enchanced by tidal forces, but nobody really knows. (NASA/Jet Propulsion Laboratory)

demolished altogether. At the time, some scientists theorized about what would have happened if Phobos had been broken up. Their calculations showed that the pieces of Phobos would form a ring of debris, orbiting Mars, and that the debris would quickly reaccumulate into a moon. Perhaps this *did* happen in the far distant past, and Phobos is really a second- or third-generation moon. These ideas of the early 1970s were never pursued, however, and Phobos remained a small curiosity until the U.S.S.R. launched an unmanned space mission to the satellite in 1988 and inspired new plans to send human beings there, as a way station to landing on Mars.

Phobos is no oddity, however. Among the smaller satellites of the solar system, it seems almost more common than not for them to have one crater nearly big enough to have shattered them. On Voyager 1's day of encounter with Saturn, November 12, 1980, the spacecraft cameras snapped some pictures of Saturn's inner moon, Mimas, showing one side of its pitted surface to be dominated by a deep crater a third the diameter of Mimas itself. One of the pictures showed Mimas with an uncanny resemblance to the Death Star space station in the first "Star Wars" movie. Then nine months later, as the second Voyager was closing in on Saturn, pictures were taken of 650-mile-diameter Tethys, another of Saturn's numerous moons. And there on Tethys was an enormous, 250-mile-diameter crater, about 40% of the diameter of the satellite itself. On the other side of Tethys, the first Voyager had already recorded the 1300-mile-long Ithaca Chasma, an immense canyon for such a small body. The canyon, stretching nearly three quarters of the way around Tethys, conceivably was a response of the small world to the enormous impact.

Scientists like Gene Shoemaker began to think about the giant craters on some of the moons of the planets. The Voyagers had shown many of the solid objects of the outer solar system to be saturated with smaller craters, just like the highland surfaces of the Moon, Mars, and Mercury. Io, Europa, and Enceladus were exceptions: The outer solar system had been subjected to cosmic bombardment, just like the inner solar system. Crater counts showed that there are relatively fewer immense crater impacts recorded in the outer solar system than on the terrestrial planets. Perhaps comets, which are more common projectiles in the outer solar system than asteroids, tend to be smaller than the larger asteroids, resulting in smaller craters. But there were plenty of craters of about 70 miles diameter and smaller, plus a few much larger ones. Evidently there

The moons of the outer planets have been heavily battered by impacts. Mimas, shown here in a Voyager photo, has one crater so large that the impact must nearly have disrupted Mimas. (NASA/Jet Propulsion Laboratory)

was a time when the frigid tranquillity of the outer solar system was dispelled by a torrential bombardment. More heretical yet, maybe the outer solar system was never tranquil.

On a large planet, the longer its surface is exposed to bombardment, the more craters form until newer ones overlap older ones and the surface becomes saturated with craters. Every once in a while, a truly enormous projectile hits, forming a huge crater or basin, and then that fresh basin floor is, itself, increasingly cratered to the point that the basin is barely recognizable. What happens if, instead of a large planet, the target of the celestial bombardment is a small satellite only 15 miles across, like Phobos, or 250 miles across, like Mimas? At first, the smaller craters build up to the point of saturation. But eventually the satellite has an excellent chance of being struck by a projectile that would threaten to form a crater larger than the body itself. The result, of course, is not a crater but a shattered and disintegrated satellite. The pieces would soon spread into a ring, circling the planet in the satellite's orbit. We would be left with a ring rather than a satellite. Could it be that the rings of Saturn are the remains of a shattered inner moon?

Saturn's splendid rings may well mark the breakup of a primordial moon. Or they may consist of material that never coagulated into a moon in the first place. Calculations, like those done for Phobos, show that usually it should take only a cosmically short time for a shattered satellite to coagulate again into a reborn moon. There is an exception: Very close to a planet—such as at the location of Saturn's rings—tidal forces and other gravitational effects prevent rapid reaccretion. While the rings probably do not mark the recent disintegration of a moon, they nevertheless remind us about the processes that undoubtedly have occurred during earlier epochs in the history of the outer planets. The large planets' powerful gravity fields tend to focus cometary projectiles, speeding them toward especially violent impacts on the satellites. Inner satellites like Mimas may be particularly vulnerable.

Gene Shoemaker has become excited about the prevalence of cratering in the outer solar system, which he attributes to a large population of comets. Planetologists can date the ages of the surface geology on these moons by counting the numbers of craters on their surfaces and by assuming that the craters form as often as they are known to do on the Moon or the Earth. But if the cometary cratering has been more intense than asteroidal cratering in the inner solar system, the surface ages could be much more recent than we had expected. That would make the inferred ages of the noncratered satellites (like Europa) younger, too, meaning that they are even more active geologically than scientists had supposed. However, without any rocks to date from the outer solar system, like the returned Apollo samples from the Moon, we cannot be sure about the rather speculative calculations and cometary tales of scientists like Shoemaker. While evidence of the giant craters is there to be seen on Tethys and Mimas, no saturnian satellite looked as though it had actually just reaccreted from a recent catastrophic breakup. So, few scientists were convinced that moons had broken apart and been regenerated in a cyclical response to cometary bombardment.

* * *

And then there was Miranda, o brave new world! This seemingly insignificant satellite remained undiscovered until 1948, when it was detected by Gerard Kuiper, the sole full-time planetary astronomer in the United States during the first decade following World War II. Like most of the larger moons of Uranus, it was named for characters in Shakespearean plays. In *The Tempest*, lovely Miranda was hidden away until awakened to the world through the agency of a mighty storm, which shipwrecked her lover-to-be on her enchanted island. In the uranian system, Miranda's namesake has a strange beauty of its own, contrasting with the familiar crater-scarred aspects of its brethren satellites. Scientists now think that the small, inner uranian moon may itself be the product of a cataclysmic history, when its precursor was smashed to pieces by a comet and hunks of the body then reassembled in jumbled disarray. Miranda remains today as a mangled moon, with some of the most diverse and oddest terrains to be found anywhere in the solar system.

"Who would have thought that Miranda would turn out to be so damn interesting?" one of the Voyager scientists asked rhetorically as the pictures came in to the Jet Propulsion Laboratory. Conditioned to expect more interesting geology on a bigger body, Voyager scientists had at first wanted to fly closer to one of the larger Uranian moons. But the rigid rules of celestial mechanics dictated that if Voyager were to proceed to Neptune as planned, it had to fly by 300-mile-wide Miranda, or nothing. So Voyager's closest photos of any moon in the outer solar system just happen to be of a small world that startled even the most jaded veteran of the decade-long Voyager mission. Many planetary scientists just cannot seem to accept that

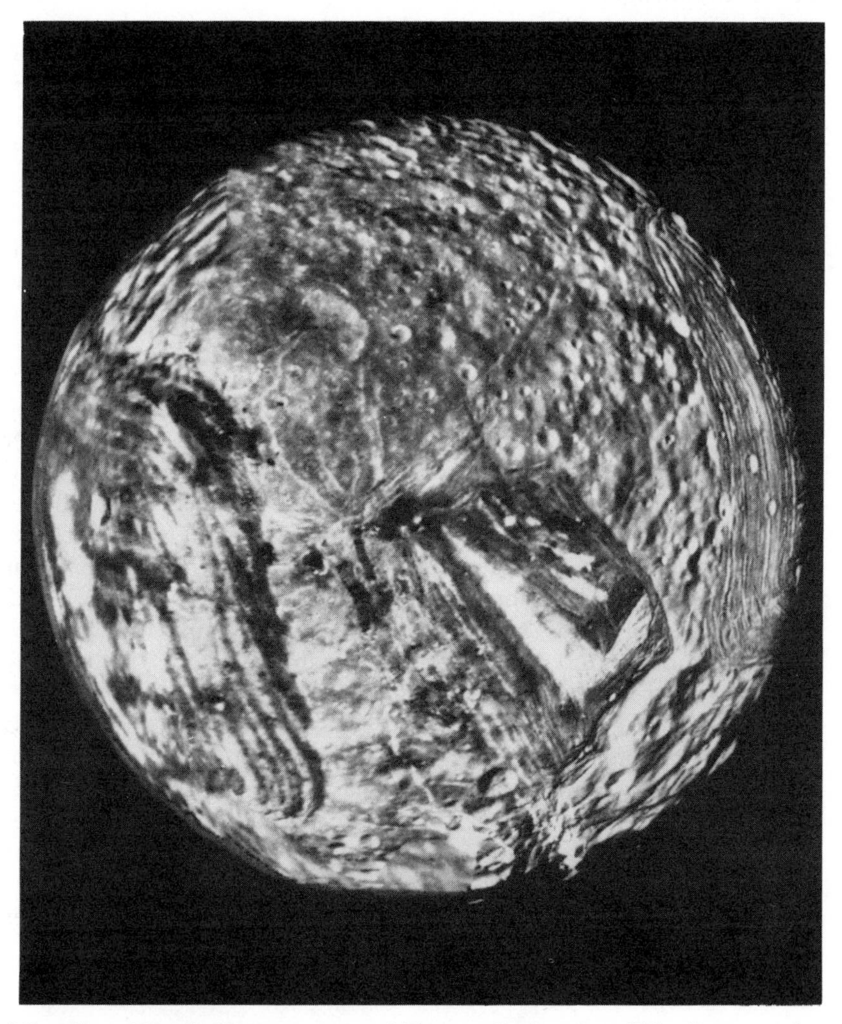

Miranda, a 300-mile-diameter moon of Uranus, displays incredibly diverse surface geology, perhaps the result of an early impact that disrupted the entire object. (NASA/ Jet Propulsion Laboratory)

small can be beautiful. "I wonder if we'll learn by the time we get to Neptune?" commented Larry Soderblom, the assistant camera-team leader for Voyager.

Even the earliest pictures of Miranda were enigmatic. From a distance, it looked as though some celestial giant had painted a big, white check mark on its surface, as if to say, "Here's the answer!" Later called "the chevron," the immense check mark remains unexplained to this day. As the hours passed and Voyager approached, Miranda came more sharply into focus. Soon, the scientists, press corps, and public alike could only gawk at pictures of the strange little world. Subsequent measurements have shown that it possesses a cliff face nearly 10 miles high. There are cratered provinces on the body, but other ovoid regions, with spiraling patterns of grooves and ridges, are nearly devoid of craters.

All the Voyager scientists could do at first was to compare Miranda with what they had seen elsewhere in the solar system. According to Soderblom, trying to explain a picture of Miranda at a Voyager press conference, "Miranda is a bizarre hybrid of valleys and layered deposits on Mars, combined with the grooved terrains of Ganymede, matched with what some have called compression faults on Mercury. So if you can imagine taking all the bizarre geologic forms in the solar system and putting them on one object, you've got it in front of you."

Several years later, scientists still cannot explain Miranda's appearance in detail. But one general idea seems at least as good as any other. It is that Miranda, in fact, is a concrete example of Gene Shoemaker's idea of a moon ripped apart by a giant comet impact, and randomly reassembled into a chaotic, jumbled configuration. Parts of Miranda's present surface may once have been on the inside of its precursor body. Who knows? Maybe there has been much subsequent readjustment, and none of the features of the original body has been preserved. In any case, the unraveling of Miranda's mysteries may teach us how beauty was born of catastrophe in a chaotic earlier epoch of uranian history.

CHAPTER 10

Chaos

The concept of "chaos" seems naturally associated with catastrophe. The most violent terrestrial catastrophes in our time—the tornadoes, earthquakes, explosions, avalanches—take the well-ordered attributes of our modern world and convert them to a chaotic jumble, with shattered buildings, ruptured utility lines, and the general random disorder we term "chaotic." The essence of the concept of chaos is unpredictability. Weather forecasters cannot predict just when an ominous tornado cloud will descend from a squall line, and the disposition of trees, homes, and vehicles in its path can be known only after the damage has been done.

In contrast to the notorious unpredictability of weather and the apparent ineptitude of forecasters is the stunning predictive success enjoyed by astronomers for centuries, even millennia. There is nothing so certain as that the Sun will rise in the east tomorrow (whether or not we can see it through the unpredictable clouds). Eclipses were being predicted by Chinese astronomers at the dawn of civilization. The return of Halley's Comet in 1986 happened right on schedule, and was just as disappointing for Earth-bound viewers as astronomers had long predicted that it would be. In fact, since the time of Newton, nothing has seemed so regular and predictable as the behavior of the planets, as they unerringly trace their prescribed courses through the heavens.

In the centuries since Newton, classical physics has provided better and better explanations of our natural world, at least of those aspects that we human beings can see with our eyes and touch with our hands. The present century has witnessed some modifications of classical physics, such as Einstein's postulate that strange things happen near the speed of light. But the voyages of Star Trek's "Enterprise" notwithstanding, relativistic physics is equivalent to

Newtonian physics as it applies to our daily lives. Physicists also have developed quantum mechanics to explain the behavior of light and matter at submicroscopic scales. Indeed, the famous "uncertainty principle" of quantum mechanics has had a major, if metaphorical, impact on 20th-century philosophy and the way we think about ourselves. But, once again, the physics of the minuscule building blocks of the universe merges into classical physics at the level where we deal with our world.

Yet the stunning success of classical physics in dealing with billiard balls, pendulums, and the motions of the planets must be balanced against its failures in coming to grips with some of the more complex, chaotic phenomena of both our natural and engineered environment. River-runners on the Colorado can only marvel as the smooth, undulating flow changes to the white-water turbulence of a rapids. How do we understand the sudden, often fatal fibrillation of a human heart that has beat like clockwork a billion times? How can we forecast the weather?

Scientists have been hoping that we would continue to make steady progress in understanding more complex phenomena, aided by the remarkable calculating power of computers. In the 1960s, meteorologists established elaborate networks of weather stations around the world to measure temperatures, pressures, and humidities: data to feed into ever more detailed computer models of atmospheric behavior. The hope was to extend our pretty reliable forecasts for tomorrow and the next day out to next week, and maybe even to the month beyond.

But it was not to be. The hopes have been dashed. There is an attribute of nature that had not been understood. It raises both barriers to, and hopes for, our eventual understanding of apparently complex, erratic phenomena in our world. In fact, unexpectedly, this chaotic behavior has its own underlying rules, generating order akin to the beauty of snowflakes and ferns. Most surprisingly of all, the newly developed science of chaos has opened our eyes to the completely unexpected role of chaotic behavior in that most well-ordered sphere of science—planetary motions. Through chaos, we have not only come to appreciate an underlying order to some of the most randomized aspects of our lives, but also uncovered unexpected possibilities for chaotic disorder in the heavens.

There is a tall, slender building on the campus of the Massachusetts Institute of Technology topped by several white ball-shaped protuberances. Inside the balls, small weather radars scan the

Boston skies, feeding data to the atmospheric scientists and meteorology students who occupy the upper floors of M.I.T.'s Building 54. A quarter century ago, Professor Edward Lorenz began some rudimentary experiments in trying to model the world's weather. He also tried to capture some fundamental characteristics of the atmospheric circulation in a few simple equations programmed to run on an early-generation computer that was available to him. He soon discovered what is now known as the "Lorenz strange attractor," opened up the new science of chaos, and—as a sidelight—proved that accurate weather forecasts will never be possible for longer than a few days, or a week or two at the most. It turns out that the tiniest whirlwind of leaves can grow into a hurricane within a matter of weeks, but just *which* whirlwind will blow itself out in a minute or two, and which will become the hurricane, is inherently unpredictable due to the specifically chaotic attributes of the mathematical equations that describe the fluid motion of the air.

Now a boyish-looking young scientist named Jack Wisdom occupies an office just a few floors below the domain where Lorenz carried out his historic studies. He has the same view as Lorenz did of white boats dotting the Charles River, although a few more skyscrapers have sprouted over in Boston. Computer technology has matured, however, and now Wisdom has access to his own small, dedicated machine to solve the equations of planetary motions extending backward toward the beginning of the solar system. Wisdom's small computer runs almost as fast as a Cray supercomputer in simulating planetary motions, and he lets it run for weeks at a time. What Jack Wisdom and his students have been discovering is that the orbits of many of the asteroids and satellites of the solar system—and the orbit of at least one planet, as well—are inherently chaotic. Jack Wisdom does not believe that Mars will suddenly depart from its nearly circular orbit and head for a terminal crash into the Earth. Far from it. In fact, his computer has reaffirmed the stability of some planetary motions. But others are in precarious orbits indeed. In fact, the continuing and potentially catastrophic rain of asteroids and meteorites onto the Earth is finally explicable—after decades of failed hypotheses—in terms of chaotic asteroidal orbits. A well-behaved asteroid may suddenly zoom toward Mars, or even the Earth. And, Jack Wisdom thinks, distant Pluto may be in a chaotic zone as well.

* * *

In 1848, astronomers discovered an eighth satellite of Saturn. It appeared as a very faint starlike object in their telescopes, barely visible due to the glare of Saturn itself. It was given the name Hyperion. The size of the point of light could not be measured directly, but Hyperion reflected enough sunlight—despite the great distance of the Saturn system from both the Sun and the Earth—that it had to be about 250 or 300 miles in diameter. Little more was learned about Hyperion until the 1970s, when astronomers first used sensitive new instruments to detect the spectral signature of water-ice on the little moon's surface.

On August 23, 1981, the Voyager 2 spacecraft was rapidly closing in on Saturn. As it passed near Hyperion, a dozen pictures were snapped from different directions. Voyager's portraits of Hyperion showed that it looks like a gigantic, pockmarked hamburger, rather more oddly shaped than most planetary satellites. In the following months, scientists at the Jet Propulsion Laboratory went about the usual tasks of analyzing the pictures. An obvious thing to measure, now that they could "see" Hyperion for the first time, was its size and shape. Evidently the "burger" was about 240 miles across and about 140 miles thick.

The next thing to measure was the rotation period and the direction of the spin axis of Hyperion. This is a little trickier, but good results had been obtained for other satellites from earlier spacecraft encounters. To assist them, the scientists used an image-processing system—a computer that could project the pictures of Hyperion on a screen and simultaneously overlay grids of latitude and longitude lines, calculated for various values of the spin period and direction. Voyager took good pictures of Hyperion for a couple of days, so Hyperion had a chance to turn on its axis quite a bit, certainly enough to be measured. In addition, photographs taken from farther out, although poorer, showed Hyperion's fluctuating brightness, as first the broad side and then the edges of the hamburger faced the approaching spacecraft.

When all the measurements were done, they simply didn't make sense. The scientists who looked at the data taken from a distance were sure that the spin period was exactly 13.1 days. But when the grids for that value of the period were superimposed on the close-in photos, it was immediately evident that Hyperion was spinning differently. In confusion, the Voyager scientists turned to ground-based astronomers, who—with great difficulty—tried to measure Hyperion's fluctuating brightness despite the glare of Saturn. They, too, failed to settle on a period.

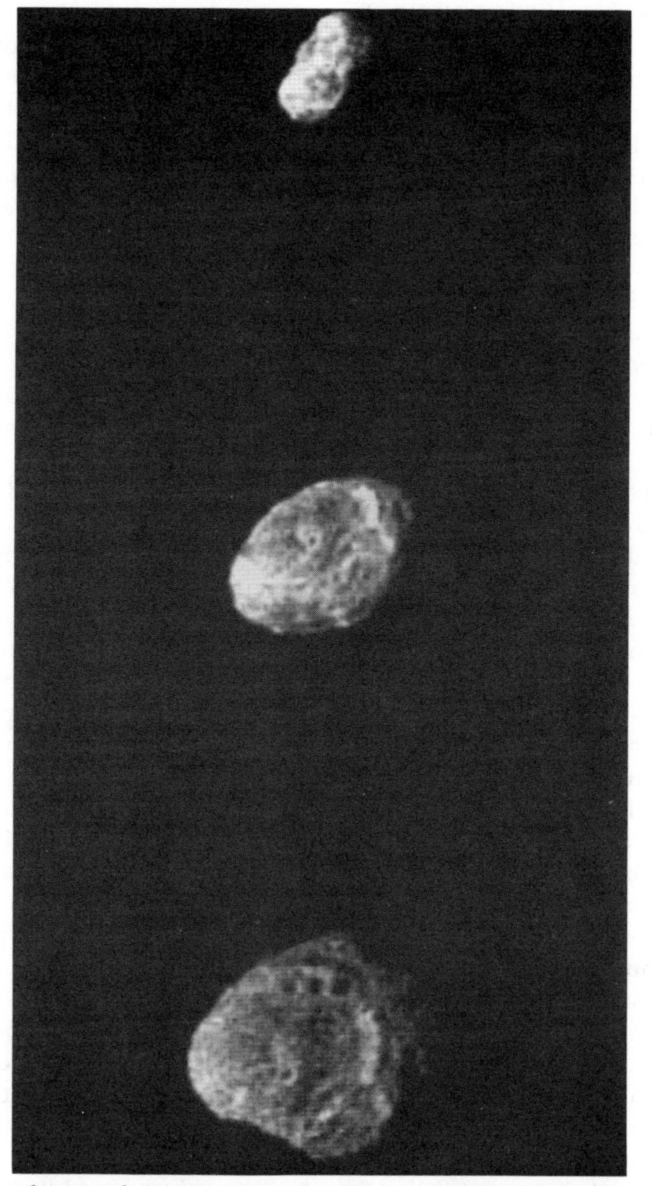

Hyperion, shown at three orientations. One of Saturn's moons, Hyperion was the first object in the solar system found to be in a chaotic rotational state. (NASA/Jet Propulsion Laboratory)

That was when Jack Wisdom, then a graduate student about to get his Ph.D. degree, appeared on the scene. His professor at the University of California at Santa Barbara was Stan Peale, one of the men who had predicted Ionian volcanism before Voyager got to Jupiter. Peale had made a prediction about Hyperion, too. He had studied how tidal forces between a planet and one of its satellites tend to slow down the spin of the satellite until it is locked into a one-to-one relationship, keeping the same face always toward the planet. That, of course, is true of our Moon, and it is also true for many of the other moons in the solar system. But Hyperion is pretty far away from Saturn and has some other orbital peculiarities. So Peale wondered whether there had been time enough for Hyperion to have become tidally locked, and he predicted that it might have gotten trapped in a special rotation period. Instead of keeping one face always toward Saturn, with its spin period exactly equaling the period of its orbit around Saturn (a Hyperian "month"), Peale thought it could now be trapped in a spin period exactly 50% faster than its "month."

Peale and Wisdom were on the lookout for Voyager's results on Hyperion. And when the data didn't make sense, they were soon ready with an answer: Hyperion spins chaotically. Now, as a tenured professor in M.I.T.'s Building 54, Jack Wisdom will be glad to show visitors his computer-made movie of Hyperion, which dramatically illustrates its lopsided, tumbling motion around Saturn. Hyperion is so misshapen and its orbit is sufficiently out of round in shape that—given the forces it feels from Saturn and Saturn's largest moon, Titan—it is in a "chaotic zone" according to Wisdom's mathematical calculations. That is, according to the science of chaos, Hyperion cannot have a stable spin or spin-axis direction, at least at this epoch in solar system history.

As encapsulated in the movie, Wisdom's theory for Hyperion's spin shows it spinning rapidly one month, then suddenly being retarded into virtually no spin at all. Sometimes its spin is aligned roughly vertically to its orbit (like most planets and moons), but a few months later the axis may be flopped down to the orbital equator. Wisdom's theory seemed to explain the inconsistency in Voyager's measurements of Hyperion, and he quickly became a celebrity in the scientific world.

At first, it appeared that Hyperion was just an oddball case of planetary chaos, but Jack Wisdom has continued to discover chaos almost everywhere he looks. From subsequent analysis of the spin dynamics for other satellites, Wisdom has demonstrated that many of them must have spent a part of their lifetimes spinning chaotically, before still further tidal action finally locked them into regular spins with the same periods as their orbital periods. Among the moons that Wisdom is sure have spun chaotically in the past are the largest martian moon, Phobos, with its strange pattern of grooves, and the weird uranian satellite Miranda. It is too soon to know for sure, but the forces acting on a chaotically spinning satellite may be part of the story necessary to explain the strange geology on those bodies' surfaces.

* * *

Why do meteorites fall from the skies? Ever since the idea was first proposed that these stones originated in the asteroid belt, the mechanism that diverts them toward Earth has been a bit of a puzzle. The asteroids, after all, are far away. Nearly all of them are beyond Mars, much farther from the Earth than the Earth is from the Sun. Asteroid orbits are a bit oblong and tilted, but their paths never approach the Earth, or even—for that matter—Mars. As we have discussed, every once in a while asteroids collide in blinding explosions. It is reasonable, at first, to guess that rock fragments are sprayed into space and some reach the Earth. But a little more thought, confirmed in laboratories, convinced scientists that it isn't so simple. Even the most violent collisions, shooting debris out at a mile per second or more, would put the fragments only into orbits near the original body's orbit. In fact, there are groups of asteroids, known as "families," that are all in similar orbits, and the accepted theory is that they are products of an asteroid collision and breakup.

Even if we were to imagine that a few collision fragments get accelerated to much higher velocities than the others, enough to bring them close to Mars or the Earth, we would run into another difficulty: Rocks given that much acceleration melt or vaporize. But many of the meteorites that fall from the skies show little or no evidence of such explosive accelerations. Their outside surfaces are melted from their fiery plunges through the Earth's atmosphere, but their lightly shocked interiors look remarkably unblemished for having traveled so far. In the 1960s, some scientists began to look elsewhere than the asteroid belt to explain where meteorites come from. Some thought the Moon was a possibility, until returned Apollo samples dashed those hopes. Others thought that meteorites might be pieces of comets, which already are on Earth-crossing orbits. In the view of those scientists, the meteorites' larger brothers and sisters—the Earth-approaching asteroids—were just dead comets.

But still other scientists, including many meteoriticists who dissect the fallen rocks in their laboratories, were loath to think that these rocks could come, via icy comets, from the outermost reaches of the solar system. Meteorites have the same general chemistry as the bulk Earth itself, so meteoriticists believed they must come from parent bodies that are at least fairly nearby, not from the cold, volatile-rich birthplace of comets. Somehow, some way, various gravitational forces must tug some asteroidal debris out of their nearly circular orbits and send meteorites our way. There are gaps in the distribution of asteroids where asteroids presumably once were but are no longer. The gaps seem to be related to the gravitational forces of Jupiter, so it was thought that perhaps Jupiter's repetitious pulls on asteroidal fragments might propel them toward Mars, where interactions with Mars might move them along toward Earth.

George Wetherill, one of the most respected planetary scientists and a man we will meet again later, tried to calculate how we might get asteroidal fragments to Earth, via such routes. He was able to show that various multistage processes (involving gravity and small collisions) might yield a few meteorites. But he could not explain the large number that actually fall; therefore, for many years Wetherill was inclined to believe that most meteorites were from comets. At scientific meetings, he disapproved of colleagues who waved their arms and said "There must be a way!" and then proceeded to assume that meteorites were pieces of asteroids without demonstrating a delivery mechanism that would work.

In the end, there was a way, thanks to chaos. Jack Wisdom went to work on the problem after he finished with Hyperion, and he discovered some remarkable facts. The asteroid belt is filled with chaotic zones. These zones are not exactly actual *places* in the asteroid belt, but regions in "parameter space" (various combinations of orbit size, shape, and orientation) that result in chaotic motion. Many zones of chaos are, indeed, associated with the Jupiter-caused gaps in the distribution of asteroids. An asteroid can orbit for hundreds of thousands of years in a perfectly regular, sensible way, and then quite suddenly its orbit can change chaotically into a cometlike, elongated path that comes near the Earth. And so we get meteorites, and without melting or vaporization and with no need for black-magical, arm-waving dynamics. Thanks to Wisdom's research, which George

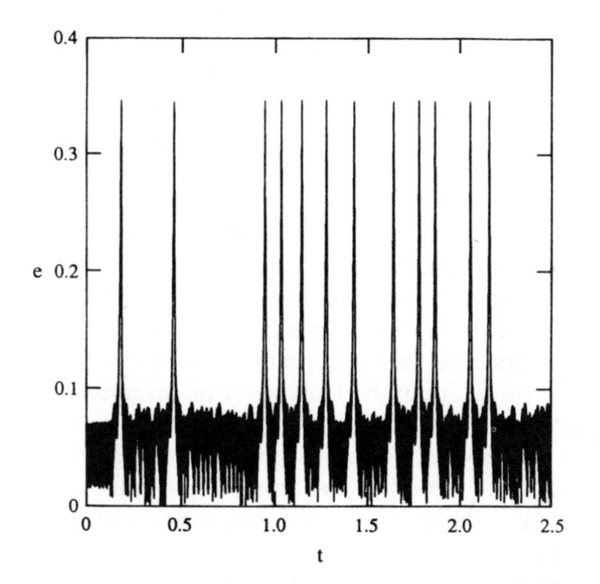

Chaotic variations in the eccentricity of an asteroid orbit over 2.5 million years, as calculated by Jack Wisdom of M.I.T. Normally the orbit is quite circular, but at irregular intervals it becomes very elongated (eccentricity greater than 0.3). (J. Wisdom)

Wetherill has now helped to promulgate through the community of meteoriticists, it is once again widely accepted that both the meteorites and many of the Earth-approaching asteroids were derived from the main asteroid belt, far beyond Mars.

* * *

Chaos may well hold further surprises for us. The science of chaos, applied to dynamical systems, now teaches us that the world of dynamics is divided into two comparable domains: the domain of regular motion, and the domain of chaos. Dynamicists have been studying regular motion in gory detail for centuries, but now there is a whole new field of chaotic motions to be explored. In a few short years, Jack Wisdom, and now his students, have discovered lots of chaos. There may be more to be found. That introduces a further element of unpredictability into the cosmos, and into our lives. For not only might a chaotically extracted asteroid fragment come plunging down onto our world, but some other as-yet-undiscovered chaotic celestial phenomenon may be waiting to reveal itself.

Chaos implies disorder, but it also has its own internal orderly aspects. From studying the orderliness of chaos, as self-contradictory as that sounds, we may learn much more that is solid and secure about our universe. For example, Jack Wisdom's ex-student Bill Tittemore, now a professor at the University of Arizona, has been studying the motions of the satellites of Uranus. By applying chaos theory, he has been able to tell us things about those satellites, and about Uranus itself, that wouldn't be knowable if the satellites weren't prone to chaotic behavior. In particular, there is now a good understanding of just what conditions lead to chaos, and to what degree of chaotic behavior. The many varieties of chaotic behavior are being cataloged and studied. If we can determine that a satellite system must have been behaving chaotically at some time in the past, even though it isn't today, we may be able to understand its present condition. For example, the grooves on Phobos may be due to that martian satellite's chaotic past.

Another useful feat is to turn chaos on its head. While we cannot predict the precise future behavior of a chaotic system, we *can* determine the *past* behavior of a chaotic system to unprecedented accuracy from measurements of its present state. For example, if astronomers can get a series of accurate measurements of Hyperion's tumbling motion, then those can be fed into the equations on a computer and *run backward*, so that we can learn precisely what Hyperion's past motion has been, which in turn would give us a handle on other unknown parameters of Hyperion, Titan, and Saturn.

One researcher suggested that Jupiter's Great Red Spot, a four-century-old storm larger than the whole planet Earth, may be a manifestation of chaotic atmospheric motions. If chaotic dynamics dominates our own planet's weather, it might apply as well to the weather on other planets. Chaotic behavior is ubiquitous in the universe, as millions of stars move about chaotically within star clusters and galaxies. Many mysteries abound in the solar system and the cosmos, and scientists have a new, and somewhat unsettling, perspective with which to try to understand them. But until chaos is more completely understood, what is as yet unlearned about our chaotic universe may threaten us with catastrophes not yet foreseen.

CHAPTER 11

Origin of the Moon

If you look at a globe of the Earth, the largest feature is the Pacific Ocean. A century ago, George Darwin (son of the evolutionist) proposed that the Moon had formed by a splitting or fissioning of our own planet. He calculated the Moon's orbital motion, and traced it back to an early time when, he believed, the Moon was only 6000 miles from the Earth's surface and the Earth was spinning rapidly, its day only 5 hours long. Darwin suggested that unstable oscillations in our rapidly spinning planet might have served to break the Moon away. Soon, someone else suggested that the immense Pacific Ocean basin represented the cavity from which the Moon came. Whatever the merit of Darwin's idea—we still think that the Moon was once much closer to the Earth, but actual spontaneous fission is no longer accepted—one lesson can be learned: Creation of the Moon from the Earth is a process of truly enormous scale. The Moon, after all, is a quarter the diameter of the Earth, although it has only 1/80th of the Earth's mass.

In the ensuing decades, the idea of the Moon somehow being ripped away from our own planet became less popular. As we will see, the idea that the Moon might be made of material derived from the Earth lives on; the chief failure of the spontaneous-fission model concerns the physical mechanism by which the Moon was derived from the Earth. As physicists thought about how the Earth might break into two pieces—for example, by spinning so fast that it simply split apart—they could not make it work according to physical laws. As so often happens in science, a seemingly attractive idea failed the test of quantitative modeling. By the middle of the 20th century, there remained no consensus about the origin of the Moon, but most scientists imagined less catastrophic ways that our planet might have gotten its satellite than the Earth spontaneously breaking in two.

One idea, and one that is still alive although in deep trouble, is that the Earth originally formed as a double planet. That really does not solve the problem, because until recently there were no rigorous theories for how our planet, or the other planets, were created. But there were some general ideas that dominated the thinking of most scientists in the middle of this century. They assumed that the Earth was formed by the accretion, or gravitational gathering together, of lots of small asteroidlike bodies, boulders, and dust that had coagulated from the great cloud of gas and dust from which the Sun and all other planets originated. Since moons accompany most of the other planets, it was easy to assume that moon formation was a natural outgrowth of planet formation. Perhaps a disk of dust and boulders surrounded a forming planet and, because of centrifugal force, they failed to fall onto the growing planet, instead gathering into one or more satellites.

However, the satellite systems of most planets are dissimilar to the Earth-Moon system. The outer planets have numerous satellites, including some very large moons that exceed the size of our own. Mars, on the other hand, has only two satellites, each so small as to seem insignificant compared with our Moon. The planet that is most similar to ours in size and distance from the Sun, Venus, has no moon at all. Compared with the other planets, only our Earth seemed to qualify as a "double planet," with the sizes of the two bodies more comparable than for any other planet. Only very recently, in the last decade or so, have we learned enough about distant Pluto to realize that it is an even better example of a double planet than the Earth–Moon: Pluto and its moon Charon are much closer to each other, and much more nearly equal in size, with diameters of 1500 and 800 miles, respectively.

Another popular theory postulates that the Moon, somehow, was created "elsewhere" and was later captured by the Earth. This "capture theory" isn't very explicit about how the Moon formed or where, but tries to address the question of why it is in orbit around the Earth. If the Moon came hurtling past the Earth, our planet's gravitational field would tug on the Moon and bend its path. If it came by slowly enough, could gravity bend the path into an orbit around our planet? The laws of physics say no, not unless something else slows the Moon down as it swings past the Earth. If the Moon coasts down into the Earth's gravity-well,* it will coast right up and out the other side unless something akin to friction slows it down. Perhaps the Moon encountered debris circling the Earth, or an extended atmosphere of the Earth, and was slowed down and captured. But calculations show such a scenario to be very improbable. It would be far more likely that some wandering would-be Moon would crash directly into the Earth than that it would be captured into orbit. And there lie some seeds that, nurtured by studies of Moon rocks returned by the Apollo astronauts, eventually grew into the idea that the Earth might have suffered giant impacts by planet-sized bodies, in which case the Moon might have been "splashed out" from the Earth itself.

Fantastic as it may seem, this view of the origin of the Moon—that it was created when a Mars-sized planet slammed into the young Earth—became the odds-on favorite in the late 1980s. Whether it will turn out to be a temporary fad, like earlier ideas about the Moon's origin, remains to be seen. But we know much more about the Moon in these post-Apollo years, and we know more about the planets and how they formed. These data have nailed the lid on the coffin of the spontaneous-fission and capture models, and raised serious doubts about the double-planet model. Perhaps, in the giant-impact model, we may finally have the answer to a question that has fascinated human beings since they first stepped out from their caves and looked up at the orb that dominates our nighttime skies.

* * *

We should hardly be surprised that our ideas about the origin of the Moon have changed as a result of the Apollo landings. In fact, whatever one believes to be the true political or economic motivations for the Apollo Project, the "official" reason for going to the Moon, as promulgated by CBS's Walter Cronkite and other shapers of public opinion, was to find out what the Moon was like, what it was made of, and—most important—how it formed. Some years after the Moon trips had ended and the news media had turned their attention elsewhere, scientists still had not settled on how the Moon had formed. But the origin of the Moon is not an easy problem to solve. It is hardly surprising, given the complexity of science, that it has taken years of additional study, and synthesis of the results from many different

*The motion of an object falling toward a central object is similar to that of a ball rolling down into a funnel, hence the analogy between a gravity field and a well.

research projects, to make much progress on solving the mystery of the Moon's origin.

Some of the most important data from Apollo concerned the ages and chemical composition of the Moon's rocks. It was quickly learned that the Moon itself is very old, as old as the Earth and the meteorites: 4.5 billion years. It was also learned that the Moon suffered incredible crater-forming bombardment during the first 600 million years of its existence, as we described earlier. We also learned that the Moon's surface is not composed of green cheese (as nobody had seriously thought) or of any common meteorite type (as many scientists had assumed). Instead, the Moon is made of its own brand of volcanic rocks, similar—but not identical—to the lava basalts that we find in volcanic regions of our own planet. Evidently the Moon was once hot and at least partly molten, so that rocks in its interior melted into magmas, and the lighter rocks flowed and floated to the surface. They may have accumulated into an ocean of magma which later cooled, forming the lunar crust, perhaps with lakes of lava.

What is the bulk composition of the Moon? That is, apart from the rocks on the surface, what is the average chemistry of the Moon throughout its interior? Lunar scientists made detailed measurements of the minor constituents of Moon rocks, and used the results in complex modeling calculations to infer the bulk composition. It turns out that the Moon does *not* have the bulk composition of the whole Earth. For one thing it virtually lacks an iron core, which makes up a major fraction of the Earth's interior. In many other ways, however, the Moon as a whole seems quite a bit like the composition of the Earth's mantle, that rocky part of our planet sandwiched between the core and the crust; the mantle makes up most of the mass of our planet. So, at least approximately, the Moon's composition was compatible with the idea that the Moon was made of material from the Earth's mantle. A few die-hard adherents to the spontaneous-fission concept briefly took heart.

Not only were most physicists unpersuaded, but they had become even more certain that fission could never have happened. And chemists analyzing the composition of Moon rocks found some small but significant differences between the Moon and the Earth's mantle. For instance, the more volatile (more easily vaporized) chemicals are deficient in the Moon compared with the Earth. The Moon might be approximately like the Earth's mantle—and indeed some chemical signatures were surprisingly alike, including the ratios of the isotopes

A complicated rock from the lunar highlands, collected by Apollo 16 astronauts. (NASA)

of oxygen—but other important signatures were inexplicably dissimilar. Like detectives trying to make sense of a bewildering myriad of clues, scientists tried some other ideas.

If the Moon wasn't formed from the Earth itself, perhaps it was formed *near the Earth* from material much like what accreted into the Earth. This is the "double-planet" scenario, and suited most scientists' prejudices, so they reexamined that idea in greater depth, buttressed by a wealth of new measurements of Moon rocks and developing ideas about planetary dynamics. The new, detailed calculations raised questions that have not been fully resolved about why the accumulating lunar material would stay in orbit around the Earth rather than spiral in to crash onto the Earth's surface.

There was an even more vexing question about the double-planet model: If the Moon was formed from material just like the material that made up the Earth, why does the Moon lack an Earth-like iron core? Not only does it lack a large core, there is little iron anywhere within the Moon. What happened to all the iron? Iron is a ubiquitous constituent of the Sun and of cosmic materials in general, including solid bodies in the solar system. Much of Mercury's interior is solid iron, other planets like Venus evidently have iron cores, and great hunks of iron continually rain down on our planet from space: the iron meteorites. Scientists began thinking about ways to separate iron from protolunar material.

* * *

In the early 1970s, geochemists John Wood and Henry Mitler of the Smithsonian Astrophysical Observatory proposed a novel approach to the formation of the Moon that seemed to account for many of the facts. First they had to think of a way of separating and removing iron from the rest of the protolunar material. What would serve as a heavenly smelter? Well, iron could be separated out by melting a large body of Earth-like composition, in which case the molten iron would sink and form a core. This is just what happened within the Earth and other planets and the way iron meteorites were made in asteroidal cores.

Suppose a small planet (several times larger than the Moon), already segregated into a rocky mantle and an iron core, came hurtling past the Earth, as in the old capture model. Wood and Mitler imagined that the Earth's tidal forces might have torn the planet apart. Their computer calculations showed that the force of the Earth's gravity would pull harder on the part of the small planet closest to the Earth than on the distant side. The differential tidal stresses would be strong enough, their calculations showed, to rupture and tear apart the interloper. They envisioned that much of the body, including the metallic core, would continue back out into interplanetary space, but that a fraction of rocky debris, derived from the mantle of the disrupted body, would be captured into orbit around the Earth. Subsequently such captured mantle rocks would gather into the Moon we know today: made of mantlelike material, but lacking in iron.

The Wood–Mitler model was attractive, but had its problems. It isn't impossible to capture a part of a body, as it is to capture a whole body, but it is improbable. Also, later calculations of the tidal physics demonstrated that it is very difficult for a body to be disrupted during the few tens of minutes it would be deep within the Earth's gravitational grip.

This "disintegrative-capture" model for our Moon's origin could have been proposed only in a post-1960s climate of scientific opinion. Before we had spacecraft pictures of near-universal bombardment of planetary surfaces, few scientists tolerated the idea of near-collisions of planets in space. The uniformitarianism of most geologists and the anathema of Immanuel Velikovsky's pseudoscientific ideas of "worlds in collision" would not easily admit ideas of a Mars-sized planet running amok around the inner solar system. Mars itself, of course, is secure in its near-circular orbit far from the Earth. But uniformitarianism was being challenged on all fronts in the 1960s and 1970s. Some scientists—especially William Hartmann, who had studied the huge lunar basins—were primed for thinking about large bodies roaming in interplanetary space, left over from the period of planetary accumulation.

It is thus not surprising that Hartmann was the first to propose that the Moon was formed when a Mars-sized planet crashed into the Earth. Hartmann's original proposal, made with his colleague and collaborator Don Davis, was presented as long ago as 1974 to a scientific meeting at Cornell University concerning planetary moons and satellites. They proposed that the whole globe of our planet was contorted by the impact, that parts of the mantle were splashed into space, and that the Moon coagulated from the lofted debris. Today, it seems inevitable that someone would soon think of the idea of a giant impact creating the Moon. It marries at least two of the earlier concepts. First is the idea of a body hurtling past the Earth, as in capture models and the Wood–Mitler disintegrative scenario. And, second, it derives protolunar material from the mantle of a geochemically segregated body (as in Wood–Mitler), but also specifically from the Earth, as advocated by George Darwin. As the concept has evolved, it now also has two elements of the double-planet scheme for forming the Moon: First, a giant impact is now viewed as a natural, if probabilistic, outcome of the formation of planets; second, the ejected Earth-mantle material forms an extended disk of debris from which the Moon eventually coagulates, which is not radically different from models which formed the lunar accretion disk in other ways.

Indeed, a couple of other scientists also thought of the giant impact model independently in the mid-1970s. But as inevitable as the idea now seems, it was not heralded as the solution at the time. Of course, the concept was new and detailed calculations of the process had not been done. Just what happens when a Mars-sized planet slams into the Earth? How does the Moon form from the ejected debris? Does the ejected debris retain the composition of the Earth's mantle or is it so hot that the more easily vaporized material is lost, perhaps explaining why the Moon has a lower content of volatile chemicals than the Earth? In fact, does the model actually explain all the small differences between lunar composition and Earth-mantle composition that have been measured? After Hartmann and the others published their views, they turned their attention to other scientific problems. And their colleagues mainly ignored the new idea. Evidently the concept needed a decade of ripening on the vine before its time had come. When a conference on the "Origin of the Moon" was held in Kona, Hawaii, in autumn 1984, the "impact-trigger" model—as it was called—was suddenly the most widely discussed hypothesis of all. It is not that intervening work had demonstrated the value of the idea. Instead, the conference stimulated people to think anew about the origin of the Moon. Looking at the physical, geological, and chemical evidence from the perspective of the mid-1980s, many scientists could see it all in a new light.

Not everyone was converted at the conference. Indeed, more scientists may still doubt the impact-trigger model than advocate it. But their objections are not as strong as objections to most of the other theories. The quibbles that remain are cautionary questions about whether further detailed research will, or will not, demonstrate the idea's compatibility with the wealth of data about the Moon. Since the conference, several research teams have been pursuing the idea. Su-

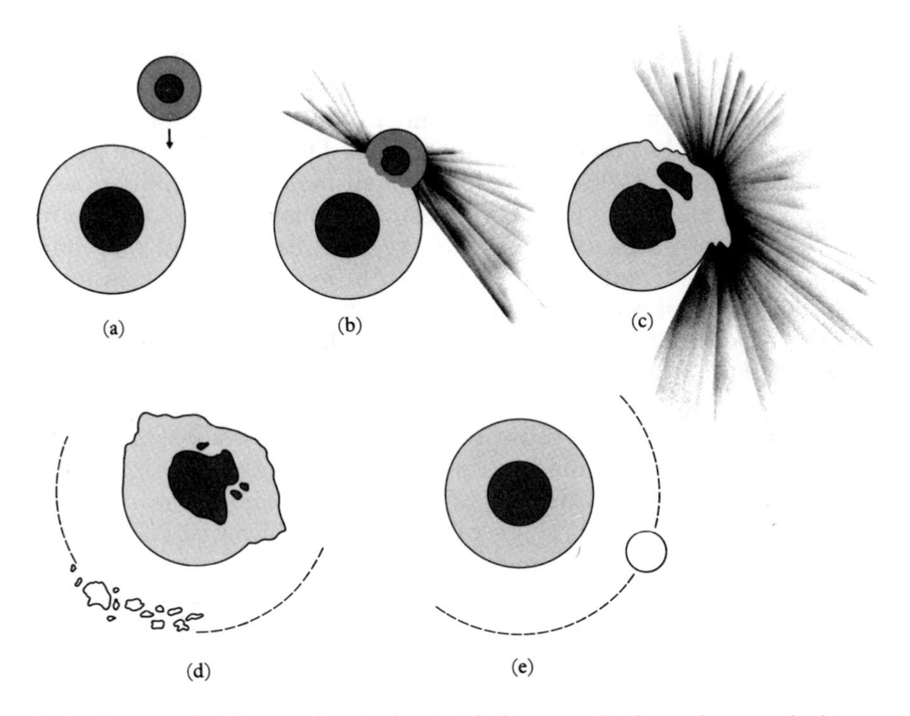

A giant impact of a Mars-sized protoplanet with the young Earth may have resulted in the formation of our Moon, as suggested in this schematic sequence based on calculations by J. Melosh of the University of Arizona. (From *The Planetary System* by D. Morrison and T. Owen, Addison-Wesley Publishing Company, 1988)

percomputer programs, written to simulate nuclear bomb blasts for the Pentagon's purposes, have been applied to this ancient and vastly more stupendous explosion. The computers trace out the trajectories of the fragments of the disrupted Earth. By modifying the equations, or changing the numbers that describe the hypothetical impact (like the mass of the Mars-sized projectile, and how far off-center it hits), the scientists can try to learn the range of possible impact circumstances that might yield our Moon. There is much work to be done, but the early results from the supercomputers look promising.

It is time that textbooks for schoolchildren be changed to suggest that our Moon may have been formed one day in the distant past when a would-be Mars-sized planet chanced to cross the Earth's orbit when the Earth itself was there. Let's suspend our skepticism and picture the story of our Moon's birth: Our planet was shaken to its roots. In the stupendous catastrophe, the Earth actually became amoeba-shaped, and an immense cavity formed extending practically to the core, as shown in the figure. During ensuing minutes an enormous amount of material rose into a cloud, which immediately collapsed into a fiery disk of vaporized rock. For many years, if our planet had been viewed by alien astronomers from another star, it would have looked like a dwarf star, with a temperature of 2000 K, glowing dimly with a reddish color. Within a century or so, the disk began coalescing into a number of proto-Moons. They in turn collided within a few years into a single orbiting body, which gradually cooled over the millennia to become our Moon.

That's our best version of how our Moon may have been born in the distant past. We are surer of what happened next: The Moon initially was molten and quite close to the Earth. Tidal forces have gradually pushed it away and locked its spin so that we see one side only. The Moon's crust solidified, and endured a half-billion years of pummeling impacts, just like the other moons and planets in our solar system. After a billion years of further cooling, the magma chambers solidified, volcanism ceased, and the Moon adopted the form we see it in today.

In our nighttime sky, the Moon rides high, perhaps a distant chunk of Mother Earth, its mare-spotted face keeping perpetual watch over its place of violent birth. That Earth-shattering calamity aeons ago, if it really happened, was the greatest catastrophe our planet has ever experienced, or is ever likely to experience again for the rest of its existence.

CHAPTER 12

Colliding Worlds

The planet Mercury is a small, elusive world, difficult to see in the twilight sky because it stays so close to the Sun. Mercury is especially interesting because its chemical composition is as different from that of the cosmos as it is possible for a large object to be. The universe (and all its stars, including our Sun) is made mostly of hydrogen and some helium. There is just a sprinkling of heavier elements, like the gases we breathe and the still heavier rock-forming elements, which have all been created within stars and supernovas. Our own non-gaseous planet is particularly unusual, being composed dominantly of rocks, with only traces of the light gases. Mercury is yet more radical: Well over half its mass is metal, chiefly the very heavy element iron. The metals we use in our lives have been refined from rocks through industrial processing like smelting. What cosmic processes have refined the planets from the nearly ubiquitous hydrogen-rich gases of the stars? And how was Mercury refined still more?

During the recent history of planetary science, Mercury has figured far out of proportion to its size in theories as to how the planets formed. Less than two decades ago, Mercury's iron richness helped inspire the theory of "equilibrium condensation," a uniformitarian scheme in which the planets were created and grew in an orderly, quiescent way. Recently, however, Mercury's strange composition has stimulated new research that points in just the opposite direction.

It now seems likely that Mercury, the Earth, and other planets may have been involved in a chaotic, violent collisional history. In this emerging perspective, scientists interpret many planetary traits to be the accidental results of stupendous early collisions and chance near-misses in an inherently unpredictable demolition derby. As we will relate, some planetary scientists now believe that Mercury is iron-rich because proto-Mercury just happened to collide with another similar-sized protoplanet; they think that the resulting gigantic impact stripped away much of Mercury's original rocky mantle and crust, leaving mainly its metallic core behind. Not only might Mercury have had a violent collisional past, but other, larger planets probably did as well, such as the Earth's Moon-forming collision we have just discussed. Such larger protoplanets were too big to be stripped to their cores, which is why Venus and Earth have a smaller bulk fraction of iron. Protoplanets smaller than Mercury may have been completely destroyed, including their cores. According to this theory, that is why Mercury is the only planet that is so iron-rich.

How could Mercury play such a pivotal role in scientists' shift from a uniformitarian to a catastrophic vision of how the terrestrial planets (Mercury, Venus, Earth, and Mars) were created? In part, it is because Mercury is truly an end-member planet—a planet extreme not only in density, but also in its proximity to the Sun. Also, the time was ripe to reconsider the innermost planet at a time when the intellectual climate was changing in planetary science.

* * *

The late Nobel Laureate Harold Urey was one of the first modern planetary scientists, and he approached the topic from his perspective as a chemist. Within a decade of the 1952 publication of his epochal book, *The Planets*, the space age commenced. But in the 1960s, most cosmochemists remained lost amid the trees of the planetary forest, studying microscopic mineral grains in meteorites fallen from the skies. Therefore, instruments to measure planetary chemistries were not included in many of the early probes to the Moon and planets. Instead, the space-probe designs were overseen by physicists and astronomers, and soon thereafter by geologists, who wished to study close-up images of the planets just as they did with aerial photographs of the Earth. Thus the "big picture" of the solar system that was emerging by the late 1960s was fashioned by astronomers, physicists, and geologists, not by cosmochemists.

That was when John Lewis arrived on the scene at the Massachusetts Institute of Technology, fresh from Urey's base at the University of California in La Jolla. A young chemist of slight build and wry wit, Lewis sought to apply fundamental principles of chemistry to the "big picture." He is now in the Southwest, writing books, raising a family, tending goats, and proselytizing for the free-enterprise utilization of space. When he arrived at M.I.T. as a young professor, John Lewis single-mindedly pursued a profound scientific question: "Why do the planets differ in bulk composition the way they do?"

For many years, astronomers had suspected that Mercury was quite dense, but it was not until two decades ago that powerful ground-based radar echoes were bounced off the surfaces of the inner planets, yielding definitive orbital positions. The planets' gravitational influences on each other could then be measured, resulting in accurate measurements of their masses and bulk densities. After correcting for compression (density increases due to gravity), scientists confirmed Mercury as the densest planet. Iron is the only cosmically abundant chemical element heavy enough to account for Mercury's density. While we cannot peer beneath Mercury's surface, geophysicists are quite sure that Mercury's rich supply of metal must be located within a huge iron core. The remaining elements in Mercury are distributed in a relatively thin, rocky shell surrounding the core.

John Lewis was fascinated by iron in the cosmos, including both the iron meteorites, like the one that caused Meteor Crater, and the immense reservoir of iron in the core of Mercury. To understand Mercury's density, he began thinking about the so-called "solar nebula," that great disk-shaped cloud of gas and dust from which, theorists believe, the Sun and all its planets were formed 4.5 billion years ago. Astrophysicists have concluded that most stars and planetary systems originate when immense interstellar clouds begin to shrink under the inexorable force of gravity, and eventually collapse into such hot, swirling nebular disks. If the inner nebula close to the forming Sun were hot enough, all the elements (including those initially in dust grains) would be in vaporized form. John Lewis wondered what chemicals would initially condense out into droplets or solid grains as the nebula later began to cool.

Lewis assumed that the nebula had the elemental composition of the Sun ("cosmic abundances," see Table 12.1), including estimates for iron that had just been recently revised. He adopted a then-accepted theoretical graph for how the nebular temperature and pressure decreased with distance from the Sun; as one would expect, the gas must have been hotter and denser close to the Sun, trailing off to wispy, cold nothingness beyond Pluto. He then incorporated the principles of chemical equilibrium—in which chemicals react with each other until everything settles down—into a computer program,

TABLE 12.1. Cosmic Abundances of the Elements[a]

Element	Symbol	Atomic number	Number of atoms per million hydrogen atoms
Hydrogen	H	1	1,000,000
Helium	He	2	68,000
Carbon	C	6	420
Nitrogen	N	7	87
Oxygen	O	8	690
Neon	Ne	10	98
Sodium	Na	11	2
Magnesium	Mg	12	40
Aluminum	Al	13	3
Silicon	Si	14	38
Sulfur	S	16	19
Argon	Ar	18	4
Calcium	Ca	20	2
Iron	Fe	26	34
Nickel	Ni	28	2

[a]These are the main raw materials available in the solar nebula for the formation of planets, satellites, comets, and asteroids.

and calculated what the composition of the condensates would be, corresponding to each temperature and pressure. He found that at the very high temperatures postulated near Mercury's current position (about 2000° F), iron grains would begin to condense from the nebular vapor, but very little rocky material would do so. Voila! Lewis's computer told him that farther out, at the distance of Venus where nebular temperatures were cooler, a lot of rocky silicate grains (made of silicon, oxygen, iron, and magnesium) ought to have condensed, as well: just enough to give Venus exactly the iron-to-silicate ratio inferred from its bulk density. Another result of John Lewis's calculations seemed to agree with our knowledge of planetary densities: Earth should be a bit denser than Venus, despite its greater distance from the Sun, because of the condensation at still lower temperatures of the mineral troilite (an iron sulfide), which is comparatively dense.

In short, John Lewis's equilibrium condensation model yielded bulk densities for the four terrestrial planets just like those that were observed. Why resort to complex models of planet formation, Lewis asked, when his simple, orderly model "does such a satisfactory job of explaining the data with far fewer arbitrary assumptions?" Scientists often love solutions that are elegantly simple, a criterion called "Occam's Razor." They forget that Nature does not always behave so simply, nor does it necessarily agree with our cultural biases about elegance and orderliness. Nevertheless, John Lewis's paper in the July 1972 issue of *Earth and Planetary Science Letters* established the benchmark for theories of planetary composition for nearly 15 years.

Condensed grains are only the beginning of the story of planetary growth. The grains must somehow have accumulated into the nine planets we know today, and several scientists were studying ways that might have happened. Evidently, as soon as grains condense out from the vapor, they should begin to rain down, just as condensed raindrops fall from the clouds. The nebular disk's gravity would cause the grains to fall toward its central plane. Soon the concentration of grains would become so high that instabilities would set in, according to calculations made by scientists in both the Soviet Union and the United States. As a result, the disk would break into countless eddies, which would quickly coagulate into billions of small asteroidlike bodies a few miles in size, each composed of the mix of condensates formed at its particular distance from the Sun. Such planetary building blocks are called "planetesimals."

Studies of the early history of the Sun indicated that the Sun would become superluminous and a powerful solar wind of atomic particles would blow away the remaining nebular gases, before they could cool and condense lower-temperature materials onto the growing planetary precursors. The little planetesimals themselves, some scientists calculated, would bump into each other and stick together. A few would quickly grow, at the expense of the others, into ever-larger protoplanets. While the final stages of growth to present-day planetary dimensions might have taken tens of millions of years, each planet's building blocks would all have come from that planet's so-called "feeding zone," the annulus extending halfway from that planet to the next inner and outer planets.

Through such a model of planets accreting from small planetesimals, John Lewis and other scientists drew the connection between the original condensed grains and the final bulk compositions of each planet. He envisioned very little mixing of materials between planetary zones, despite the fact that today Earth is struck by asteroids and comets from all over the solar system. In the 1970s, planetary theorists could understand how asteroids and comets eventually got stirred up into their unruly orbits; they realized that asteroidal and cometary collisions with planets might have added late, thin veneers of exotic materials onto Mercury, the Earth, and other planets. But bulk composition could hardly be affected by a veneer. Instead, Lewis and others concluded that the compositions of the planets were explained as the inevitable result of chemical equilibrium condensation at different distances from the Sun, and an orderly, uniformitarian accumulation process that involved very little interplanetary mixing.

* * *

Today many scientists have changed their perspectives and are beginning to think that Mercury is a fragment from the stupendous collisional destruction of a protoplanet. Moreover, serious scientific papers have been published that show that proto-Mercury (the precursor planet before this collision) may have wandered around the inner solar system during its formative stages, maybe as far away from the Sun as the orbit of Mars. Not only that, but the Earth and Venus were wandering around as well, amid numerous Mars-sized planets that have since been destroyed, or have been flung out of the solar system altogether by close encounters with their siblings. How have we come to this radically different vision?

Scientific research advances in complex ways, and there are

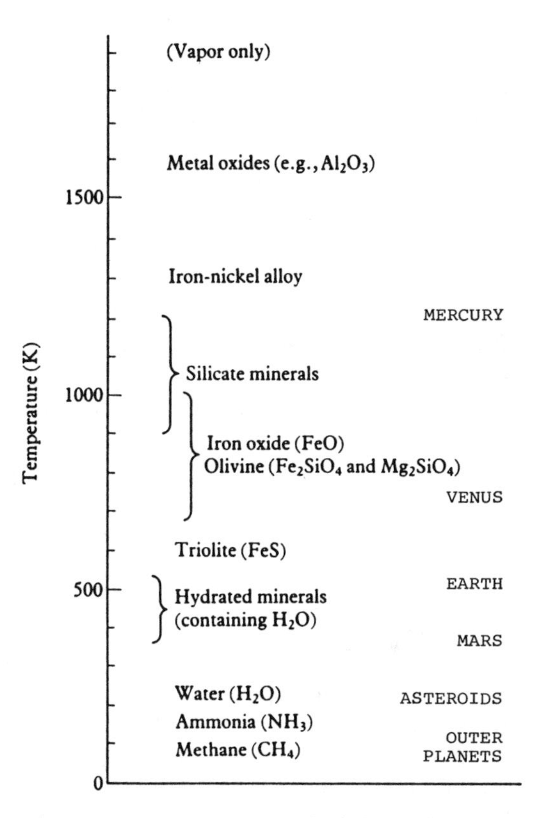

The chemical condensation sequence in the solar nebula, showing the primary chemicals that would be produced from the equilibrium cooling of a solar nebula of cosmic composition. The bulk compositions of the planets correspond with decreasing temperatures with distance from the Sun, but only very approximately. (From *Realm of the Universe*, 4th ed., by G. Abell, D. Morrison, and S. Wolff, Saunders College Publishing, 1988)

probably as many different versions of how the thinking has changed as there are descriptions by blind men of the fabled elephant. As we have told, one path involved new thinking about the origin of the Moon. Another was inspired by discoveries about Mercury made when a Mariner spacecraft flew past the small planet in 1974, two years after John Lewis's paper was published. Mariner's trajectory brought it back to Mercury two more times, and by the mid-70s a wealth of data had been returned and digested by the Mariner experimenters. Designed and developed by physicists and geologists, who overlooked the importance of chemistry, Mariner 10 was not equipped with instruments sensitive to planetary composition. But its data nevertheless reinforced the view of Mercury as an iron-rich planet. And Mariner made some other discoveries that undercut some of John Lewis's assumptions.

The spacecraft found that a magnetic field is generated within this small, slowly rotating planet. Geophysicists have long had a general understanding of how our own planet's magnetic field is generated, and it involves dynamo action in Earth's molten iron core. Thus the unexpected magnetic field means that Mercury not only has an iron core, but the core must be at least partly molten, as well. What does that have to do with John Lewis's model? As it turns out, a lot, for—as we will see—that implies that Mercury cannot be made solely of grains condensed, in equilibrium, near its orbit. Geophysicists were perplexed about Mariner 10's results, because Mercury's iron core should have frozen solid long ago. Mercury is too small to store heat for long, and it takes very high temperatures to keep iron molten. The solution to the puzzle is that there must be impurities like sulfur in Mercury's core, as there are in the Earth's core. They would lower the melting temperature and keep Mercury's core molten despite the planet's cooler internal temperature. However, neither sulfur nor any other postulated impurities could have condensed anywhere near Mercury's orbit according to Lewis's calculations. Evidently the equilibrium-condensation model was seriously flawed.

George Wetherill was the man who eventually proposed an alternative story about how this iron-rich planet came to be. A leading geophysicist, meteoriticist, and planetary physicist for several decades, Wetherill left his U.C.L.A. professorship in the 1970s to join the Carnegie Institution in Washington, D.C. Originally he had made a name for himself as a geochronologist, dating ancient rocks and meteorites. But his most important research at Carnegie has been in the very different field of planetary orbital dynamics, such as his calculations about how to derive meteorite fragments from the asteroids, discussed earlier. Wetherill can now program his theories and calculations into ever more powerful and versatile computers, permitting him to trace the gravitational interactions among a swarm of bodies of different masses. Wetherill justifiably fears that some scientists will be seduced by computational gymnastics and deluded into believing that computerized models of planetesimal growth somehow recapitulate the real processes of the actual early solar system. To be sure, models are sometimes twisted to accommodate a computer's shortcomings, some of the numbers used in the programs may have questionable bases in theory or experiment, and programming "bugs" can beget erroneous results. But as researchers have gained more experience in computer modeling over the past decade, a consensus is beginning to emerge that Wetherill's own calculations may have profound implications for our understanding of early solar-system history.

Just as Wetherill was developing the computer capability to study the simultaneous growth of several planets (instead of just one at a time), the allure of Mercury beckoned once again. It had been a decade since the last post-Mariner 10 conference, and several scientists thought it was time to take stock once more. As was his habit, George Wetherill complained of there being too many meetings, and he expressed little initial support for a Mercury Conference. But by the time more than 100 Mercury enthusiasts gathered in Tucson in August 1986 (in midsummer, Tucson seemed an appropriate venue for discussing a sizzling planet), George Wetherill was there and he offered some of the most provocative results about the small planet.

His new calculations showed that Mercury could, very plausibly, be regarded as a large iron-rich fragmental remnant from an immense collision. Moreover, in Wetherill's simulations, the peripatetic wanderings of all the developing terrestrial planets cause major interzone mixing of materials throughout the inner solar system. To be sure, the objects in Wetherill's computer runs that ultimately accumulate into Mars tend to be made of a somewhat greater fraction of low-temperature condensates than the ones that become Mercury. That is but a pale reflection, however, of the original concept that planetary properties were practically preordained by their distances from the

George Wetherill of the Carnegie Institution of Washington. (Photo by Clark Chapman)

Sun. Wetherill, in effect, tossed the whole concept of a narrow "feeding zone" on the garbage heap, and relegated equilibrium condensation to a secondary role at best.

So how, precisely, did Mercury form, according to George Wetherill? The surprising answer is that nobody knows, or ever can know, not even George himself. Inherent in his theory is that collisions responsible for fragmenting and accumulating the planets have been random and unpredictable. In one of Wetherill's computer runs, about two dozen simulated bodies, each one third the mass of the Moon, coagulated to form a rocky and metallic proto-Mercury almost twice the mass of Mercury today. Intermediate output from the computer indicates that most of the bodies originally came from near Mercury's present orbit, *or* from beyond the Earth's. After 6 million years, Wetherill's simulated proto-Mercury was at the Earth's distance from the Sun, when it suffered an enormous collision that deposited so much heat that proto-Mercury melted throughout. About 10 million years later, when it was located halfway to Mars, it was struck at 12 miles/sec by a body larger than half its size. That stripped it nearly down to its molten core. The hypothetical proto-Mercury changed little in mass over the subsequent 18 million years, while its billiard-ball trajectory finally placed it in its present, more circular orbit, close to the Sun.

We believe all of this *could* have happened. But that dramatic history is just one typical example of George Wetherill's computerized experiments. By the luck of the draw, the collisions and orbital wanderings could have been entirely different. That is the essence of random, or "stochastic," processes, which don't sit well with many scientists who were raised with the traditional view that the greatest triumphs of science involved its ability to *predict*. How much more elegant to have a theory of planetary origins in which each of the planets *necessarily* develops to be just like we know it. Many scientists are skeptical of theories which propose that the Earth has a Moon, and Venus does not, because our planet "just happened" to get hit by a Moon-forming impact, while Venus did not. "That's ad hoc," they cry, meaning that it is easier for a theorist to invent a huge impact to account for the Moon than to work out the details of an orderly process that could result inevitably in our Moon being formed. Such scientists are emotionally unsatisfied that the answer to this ancient problem could lie in an inherently unpredictable event.

But, as William Hartmann wrote in a paper entitled "Stochastic Is Not Equal to Ad Hoc," we have learned that isolated random events are just as much part of the natural universe as are smooth, continuous processes. Random events have statistically predictable chances of occurring, even though individual events cannot be predicted deterministically.

The major difference between John Lewis's original view of accretion and George Wetherill's more recent scenario can be thought of in terms of the size distribution of late-stage planetesimals and protoplanets. As we have discussed earlier, a size distribution measures how many smaller bodies there are for each body of a given size. The size distribution of ripe apples on a tree, in which few apples are less than half the size of the biggest one, is very different from the size distribution of pebbles in a stream bed. In the latter case, there may be hundreds of small pebbles for each boulder and literally countless grains of sand for each small pebble; such distributions weighted toward the smallest sizes are signatures of fragmentation and other erosional processes, whether it be pebbles tumbling in a stream or asteroids colliding in space. The statistical predictability of behavior for countless grains of sand is much more reliable than for a few large boulders, just as a thousand coin tosses will come up very nearly 50% heads and 50% tails, while three coin tosses may well come up all heads, or all tails.

The most extreme version of smooth, predictable equilibrium condensation would be if each planet had accreted directly (and quickly) from the trillions upon trillions of individual condensed grains. With the little grains continually bumping into each other in their almost perfectly circular orbits around the Sun, there would have been no chance for orbital wanderings. The difference in Wetherill's scenario is that he believes planets were mostly grown from just a few dozen, or possibly a few hundred, large bodies. If that is actually how it happened, chance events were inevitable as these few bodies interacted, and deterministic predictions of planetary traits are impossible. This view is just another step taken by 20th-century science—a famous one was the development in physics of quantum mechanics and the uncertainty principle—away from the 19th-century positivist hopes that the universe might run in precisely predictable ways.

A central question in cosmogony, the study of the origin of planets, concerns the actual nature of the size distribution that has led to planetary accumulation. Everyone has long agreed that originally all solid matter in the solar nebula was in the form of grains, no larger than the particles in cigarette smoke. Most were condensed from the cooling nebular vapors, although some never-vaporized interstellar grains may have been there, too, depending on how hot the nebula became before it started to cool. Obviously today nearly all the mass in the solar system outside the Sun is contained in just nine planets. For example, the combined mass of all the asteroids is only a small percentage of the mass of the Moon, while the Moon and other large satellites are each only a small percentage of the mass of the Earth. The important question for planetary growth is, what were the size distributions like during the middle of the accretion process, between dust and final planet? Theoreticians have considered this question since it was first raised quantitatively in the 1950s by Soviet cosmogonists. The processes in the middle stages of planetary accretion are particularly difficult to model in the computer, so the answer isn't yet in as to whether George Wetherill's simulation of the planet-growth endgame (which *began*, remember, with just a few dozen or a few hundred bodies) will prove to be compatible with these earlier middle stages still under study.

By now, however, the observationalists and experimentalists have turned the tables on the theorists and are out front in challenging equilibrium condensation. Mercury's iron core is an undisputed fact, and even John Lewis himself has given up imagining that an entire planet could form of metal, since rocky silicate minerals would condense onto it if the temperature dropped by a mere 25° F, as it surely must have during the millions of years it takes to accrete a planet. And, as we discussed before, our own Moon reminds us that no uniformitarian theory for lunar origin has explained the chemistry of Moon rocks 20 years after they were returned by the Apollo astronauts.

Looking farther into space, we see that Uranus is tipped on its axis, Venus hardly spins (perhaps due to a giant impact), and immense crater basins are visible on all the ancient solid surfaces in the solar system. Such space-age evidence of colliding worlds and immense projectiles won't go away. It is now up to the theorists and computer modelers to continue their calculations and prove whether or not George Wetherill's provocative final chapter is consistent with earlier chapters in the Great Mystery of Planetary Origin. Spacecraft studies of the planets, especially of Mercury, are raising challenges to our previous conceptions and are stimulating the theoreticians to explain them. George Wetherill has shown that a catastrophist theory for the early history of the solar system in fact makes sense.

CHAPTER 13

Catastrophism Gone Wild: The Case of Immanuel Velikovsky

Several decades before most scientists had become aware of the accumulating evidence for violent and catastrophic events in the solar system, a remarkable book broke upon the literary scene in the United States. In 1950, the prestigious Macmillan Press published *Worlds in Collision*, by a Russian-born Doctor of Medicine named Immanuel Velikovsky. Preceded by a highly favorable review in *Harper's Magazine*, this book was an instant success, selling 55,000 copies in the first six months and earning Dr. Velikovsky widespread fame and a body of enthusiastic supporters. He claimed, long before it was fashionable to do so, that there have been collisions and near-collisions among the planets, and that the history of the Earth has been marked by violent events of cosmic origin. Subsequently, Velikovsky expanded upon his catastrophist thesis in two additional books: *Ages in Chaos* and *Earth in Upheaval*.

Is Velikovsky the prophet of the new catastrophism, a person who overcame the prejudices of his time and leapfrogged into a new conception of geology and astronomy? Some people claim so, but we are not among them. The fact is that Velikovsky (who died in 1979) was almost entirely wrong in his ideas about Earth history. His work was an aberration, one of those strange byways into which intellectual history is sometimes diverted for a time. It is instructive to ask where Velikovsky went wrong and to see why his ideas represented a dead end, even though not too many years later many mainstream scientists were beginning to appreciate the true evidence for catastrophism on a planetary scale.

* * *

Unlike the other scientists discussed in this book, Velikovsky did not base his theory of planetary collisions on new evidence from geology or astronomy. He made no observations, did no experiments, and carried out no calculations. Rather, he was motivated to find a natural explanation for a variety of myths and ancient traditions, cutting across many cultures, that recounted natural and supernatural catastrophes experienced millennia ago. Prominent among these were the Judaic stories of the Great Flood and Israel's flight from Egypt across the Red Sea. Velikovsky attempted to find parallels in these diverse stories, and to suggest that the events described were not only real, but global in their effects. This is the main thrust of his argument, which is based primarily on studies of ancient texts, especially Middle Eastern sources from the first and second millennia B.C.

What could have caused global catastrophes of a magnitude sufficient to create a worldwide flood, generate plagues, part the Red Sea, and even (in the story of Joshua before Jericho) halt the Earth in its rotation so that "the Sun stood still upon Gibeon, and the Moon in the Valley of Ajalon"? Velikovsky thought that such events could happen only as a result of near-collisions of other planets with the Earth, involving both gravitational and electrical interactions. For the cause of the Great Flood, he briefly speculated about a possible splitting or fissioning of Jupiter from Saturn, a speculation never described in detail. But for the events of the first and second millennia B.C. he was more specific.

Velikovsky's reading of ancient myths convinced him that Venus had first appeared on the celestial scene only about 3500 years ago, traveling in an elongated (cometary) orbit, and that it nearly collided with the Earth on several occasions, in the process stopping and reversing the Earth's daily rotation on its axis and generating widespread catastrophes such as earthquakes, tidal waves, volcanic eruptions, and electrical discharges. These interactions perturbed the orbits of both Earth and Venus, providing a few centuries of relative peace; then the wayward Venus encountered Mars, which was thrown into an Earth-crossing orbit, and the disasters began again. There followed several close encounters between Mars and Earth, generating a series of global catastrophes only a little less violent than those attributed to Venus. Finally, about the 8th century B.C., Mars and Venus managed to settle into their present nearly circular orbits, and planetary collisions ceased. Most of Velikovsky's books are devoted to describing the evidence for these planetary encounters and the resulting worldwide destruction they caused.

* * *

From the beginning, outraged scientists—especially astronomers—criticized these books and castigated a gullible public for giving any credence to such obvious nonsense. They pointed out that planets do not go around changing their orbits and colliding with each other, and that Velikovsky's efforts to appeal to electromagnetic effects as possible explanations for such bizarre planetary behavior made no physical sense. However, these criticisms had little effect on Velikovsky's book sales, or on his followers, who noted caustically that modern astronomers had not been around several thousand years ago to measure the orbits of the planets. The Velikovsky argument was that the ancient myths and religious records said that these violent events actually took place. The evidence was there, in the form of the writings. If the theories of modern physics and astronomy were not consistent with such celestial events, then the astronomers and physicists had better modify their theories to conform to these "facts."

Granting a certain logic to these arguments, the questions raised by Velikovsky probably should have been debated primarily among the archaeologists, anthropologists, and students of ancient languages who could best comment on the correctness of the "facts" he derived from literary and historical sources. Such was not the case, however. Although Velikovsky's ideas seemed bizarre to these scholars as well, they have been less interested in joining the debate. But some have noted that Velikovsky was highly selective in his choice of quotations from ancient writings, and he had often used translations from older sources long since revised by modern scholars. When these texts are examined in detail, they frequently contradict Velikovsky's interpretation.

Even more fundamental is the problem of literalism. The essence of Velikovsky's method is to interpret ancient myths literally. Thus, for example, when the Homeric poems refer to combat between the Greek gods Zeus (Roman Jupiter) and Ares (Roman Mars), for Velikovsky this is a record of an astronomical event, involving interactions between the planets Jupiter and Mars. We hardly need to add that his is very much a minority opinion among classical scholars.

Another difficulty with Velikovsky's approach concerns the timing of his catastrophes. It is true that many ancient legends describe violent natural events such as floods and earthquakes, but it is not at all clear that these represent descriptions of global phenomena that happened everywhere at the same time. Certainly our modern experience teaches us that floods and earthquakes are confined to rather restricted areas. To make his case for cosmic causes, Velikovsky had first to establish the worldwide synchronism of events described in his sources. Doing so placed him in direct confrontation with the evidence of archaeology, especially as the concurrent development of carbon-14 radiodating methods began for the first time to establish firm dates for a number of events in prehistory, such as the explosive eruption of the Greek island of Thera (also known as Santorini) that did widespread damage in the Eastern Mediterranean between 1600 and 1700 B.C.

In order for his scenario to hold together, Velikovsky had to argue for a Mid-Eastern chronology that was completely out of step with prevailing understanding in the field. He challenged Egyptologists, in particular, by deleting four dynasties from the historical record and moving the famous Eighteenth Dynasty from the 16th century B.C. to the 10th (where Pharaoh Hatsepsut, usually dated 1503–1482 B.C., was equated with the biblical Queen of Sheba, in the reign of Solomon). In his scheme Rameses II, usually believed to be the Pharaoh of the Exodus, became a contemporary of the kings of Judah and Israel some 600 years later. His chronology wreaked similar havoc with the accepted notions of classical history, displacing the siege of Troy by the Mycenaean Greeks to the period of the Argolid tyrants. None of this speculation is taken seriously by either historians or archaeologists.

Although these problems with Velikovsky's sources and basic interpretation should have discredited his ideas from the start, most public discussion (and argument) has centered instead on his astronomy. The idea of worlds in collision, of dramatic celestial events intimately connected with human religious history, had widespread appeal. As Carl Sagan has often noted, we seek cosmic connections, evidence that our human existence is related to the large-scale forces of the universe. Velikovsky provided just such a vision. Nearly 40 years after the publication of *Worlds in Collision*, his books are still being reprinted, books and journals are published discussing him and his theories, and debates about the accuracy of his astronomical ideas continue to spring up in the letters pages of *The New York Times* and elsewhere.

* * *

Let us look at some of the major astronomical debates that have fueled the Velikovsky controversy, trying to give his ideas the dispassionate criticism we would afford any new scientific hypothesis. By a long process of trial and error, scientists have developed a technique for testing new ideas called the scientific method. The process works something like this: First, we expect a theory to be consistent with the vast body of preexisting knowledge and not to contradict the results of established experiments. (Velikovsky meets this test, in a way, by asserting that different forces acted on the planets in the past, forces that can no longer be experimentally verified.) Second, we ask that the new theory be logical and internally consistent. Where appropriate, we also expect it to be quantitative and mathematical. At the least, it must not violate quantitative reasoning or contradict experimental results. Finally, and most important, we examine a new theory to see what predictions it makes that differ from those of alternative theories, and seek to "test" the new theory by carrying out observations or experiments in these areas. Thus, for example, Einstein's theory of general relativity when published in 1917 made predictions about the gravitational bending of starlight that could be tested only during an eclipse of the Sun. Expeditions were organized to carry out these tests at the next opportunity, in 1919. When the eclipse measurements showed that relativity theory had made the correct prediction (and that the previous Newtonian theory of gravitation was, in this case, wrong), Einstein's great contribution was recognized. In a similar way, Velikovsky and his supporters have repeatedly claimed that he made predictions about the nature of the planets that have since been demonstrated. We will consider three examples.

Venus plays a central role in Velikovsky's scenario, since he asserts that it was ejected from Jupiter about 3500 years ago as an incandescent mass of gas and subsequently interacted violently with both Earth and Mars. Velikovsky asserted that Venus would still be hot (although he didn't indicate how hot) and that it would be seen to be cooling down at a rate of several degrees per year, in accord with its young age. These predictions were made in 1950, at a time when we knew very little about the physical nature of Venus. Ten years later radio astronomers discovered that indeed the surface of Venus was very hot—about 800° F—a fact that was verified by the first interplanetary spacecraft, Mariner 2, and by the subsequent series of Soviet Venera surface landers. No astronomers had expected this situation or predicted it in their theories.

What a success for Velikovsky! But there were problems. Even though the surface was hot, the cloud tops of Venus were cool, and there was no indication that Venus as a whole was losing excess heat to space. In fact, astronomers soon suggested that the surface of Venus was hot because of the blanketing action of an atmospheric greenhouse effect. Later computer models demonstrated that this trapping of heat by the greenhouse effect could raise the temperature of the surface and lower atmosphere to the levels observed. In addition, there was no evidence of any decline over time in the temperature of the planet. Thus the new data actually contradicted Velikovsky's basic thesis, that Venus was rapidly cooling down from a recently incandescent state.

Another Velikovsky prediction about Venus concerned the composition of its clouds. He attributed the manna that fed the Israelites in the desert and the naphtha also reported in the Old Testament to hydrocarbons derived from the atmosphere of Venus, and he asserted that "the presence of hydrocarbon gases and dust in the cloud envelope of Venus would constitute a crucial test" for his theory. When Mariner 2 flew past Venus in 1962, this first interplanetary spacecraft carried only five instruments, and none of them was designed to illuminate the question of the composition of the clouds. But at the press conference held by NASA following the flyby, some reporters asked one of the scientists, Lou Kaplan of the Jet Propulsion Laboratory (JPL), what material the clouds were made of. Kaplan replied that Mariner hadn't been able to determine the composition, but that based on other evidence he personally thought the clouds might be composed of hydrocarbons.

The resulting press story was headlined: "Hydrocarbon Clouds Found on Venus by Mariner 2." Unfortunately, the error was compounded when the JPL Public Information Office repeated the story, which eventually even found its way into official NASA documents on the Mariner mission. Five years later, sensitive telescopic measurements showed that hydrocarbon gases and clouds are both absent on Venus, and the clouds were eventually identified as composed of sulfuric acid droplets. Still, some Velikovsky supporters assert today

Surface of Venus as photographed from two of the Soviet Venera landers. Writing before we knew anything about the Venus surface, Velikovsky asserted that Venus was an incandescent comet ejected by Jupiter just a few thousand years ago.

that his prediction of hydrocarbons was confirmed by Mariner 2, and that the story was subsequently covered up by the government as part of a conspiracy designed to discredit Velikovsky.

Our third example of a planetary test of Velikovsky's hypothesis concerns the surface of the Moon. What was happening to our satellite at the times Venus and Mars were interacting so dramatically with the Earth? According to Velikovsky, the Moon was even more roughly treated, as "the Moon's surface flowed with lava and bubbled into great circular formations . . . [generating features that] bespeak the dreadful devastations, even death itself, that interplanetary contacts can leave in their wake." The above statement was made in *Worlds in Collision*, and on the day before the first Apollo lunar landing he repeated his concern in a signed editorial in *The New York Times*, where he warned of frequent moonquakes and high levels of radioactivity, and stated, "I maintain that less than 3000 years ago the Moon's surface was repeatedly molten and its surface bubbled." Yet, as we all know, there was no excess radioactivity, and no moonquakes were felt—indeed, it turns out that the Moon is more than a thousand times less active seismically than the Earth. Even more to the point, precise age measurements of returned lunar samples revealed that even the youngest lunar lava flows solidified more than 3 billion years in the past, in contradiction to Velikovsky's prediction of widespread lunar melting 3000 years ago. Except for an occasional crater-forming impact, our satellite has remained remarkably unchanged for the past several billion years.

* * *

When you think about it, there is something a little strange about turning to the limited evidence from other planets to confirm or disprove a theory of violent events on Earth. Surely the most straightforward tests of Velikovsky's theory are to be found right here, on our own planet. In 1950, when *Worlds in Collision* was published, geology was suffering from the uncertainties and contradictions that preceded the development of the theory of plate tectonics, so perhaps we can forgive Velikovsky for some of his errors in discussing the evidence from geology. Since that time, however, our understanding of the Earth and the other planets has come a long way, and ideas that might have seemed possible 40 years ago are no longer tenable.

Suppose the Earth really had been subject to the great earthquakes, tidal waves, volcanic eruptions, and prolonged atmospheric disturbances suggested by Velikovsky. Surely there would be ample indications of such events in the geological record. After all, in the early 1980s scientists were able to identify and measure the evidence of a catastrophic event that took place 65 million years ago—the asteroidal impact that ended the Cretaceous period. Velikovsky's near-collisions, which he alleged had stopped the Earth in its rotation, would seem to be of at least comparable magnitude, and yet they are supposed to have happened in historic times. Unfortunately for Velikovsky's followers, there is nothing in the geological record to support his case.

An especially clear indication that the Velikovskian global disturbances are fictitious is provided by the Bristlecone pines, those remarkably longlived trees in the arid mountains of the California–Nevada border. The oldest of these trees have been alive for more than 4000 years, as we can determine by counting their annual growth rings. When account is also taken of dead trees, the tree-ring record goes back at least to 3435 B.C. These rings preserve a continuous record of the local climate, since the amount of growth in each ring depends on conditions (sunlight, temperature, rainfall) in that year. While these records indicate fascinating cycles of rainfall and temperature that are of great interest for the study of the Earth's climate, they reveal nothing unusual at the times of Velikovsky's supposed catastrophes.

Even stronger evidence has been derived recently from cores bored into the Greenland ice sheet. Each year a new layer of ice is deposited, with the annual layers varying in thickness from about 1 to 7 inches. These ice cores provide a continuous annual record that extends back more than 10,000 years, preserving information on temperature and precipitation as well as layers of dust corresponding to major volcanic explosions. For example, these cores clearly show the dust signature of the Thera eruption in the Mediterranean, which can now be dated specifically to 1645 B.C., with an uncertainty of a few years at most. (In his revised Mid-Eastern chronology, Velikovsky assigned the Thera eruption to 950 B.C.) And as with the tree rings, the ice cores show no indication of disturbances at the times required by Velikovsky for his near-collisions with Venus and Mars. Any credibility that might have still accrued to Velikovsky's theory of cosmic catastrophes a decade ago has by now clearly been laid to rest. The evidence is in, and Velikovsky was wrong.

* * *

What is one to make of this strange story? We might draw upon it to think about the nature of scientific evidence and the widespread misunderstanding of scientific methods displayed on both sides of the issue. The essence of the modern scientific method is, as we have noted, the testing of predictions by acquiring new evidence that distinguishes between competing theories. But what constitutes proof or disproof? What is one to think when the pro-Velikovsky faction, for example, cites the unexpected discovery that Venus is hot as proof of the correctness of his ideas, while others argue that the Greenland ice cores prove him wrong? Which kind of evidence is valid?

The answer can be found in the discipline of formal logic, which was invented by the ancient Greeks and perfected by medieval monks in Europe. This study tells us that it is much easier to disprove a statement than to demonstrate that it is correct. Consider, as an example, the syllogism:

> If A is true, so is B.
> B is false.
> Therefore, A is also false.

This statement is easier to understand when we substitute a simple example:

> If the Sun has just risen, it is daylight.
> It is not daylight (it is dark).
> Therefore, the Sun has not yet risen.

This ancient syllogism has the same logical form as some of the critical arguments raised by scientists against the Velikovsky theory:

> If the Earth and Venus nearly collided in 1500 B.C., there should be a global layer of ash deposited by the ensuing volcanism.
> There is no layer of ash in the Greenland ice cores at that time.
> Therefore, there was no global volcanism and no near-collision with Venus.

Or another example:

> If Venus and Mars disrupted the Moon within the past few thousand years, there must have been widespread melting and flows of lava on the Moon's surface.
> Apollo showed that no significant volcanism has taken place on the Moon for billions of years.
> Therefore, the Moon was not disrupted by Venus and Mars in historic times.

Each of the above statements is correct, and each is capable of disproving, or "falsifying," the basic thesis of Velikovsky. Unfortunately, the reverse is not the case; just because an observational or experimental result agrees with a theory, we cannot be confident that a theory is proven. After all, there might be many other theories that are also consistent with the data. Successful prediction is important for a theory and it may boost our confidence that we are on the right track, but it can never constitute real proof in any absolute sense.

Science and logic both teach us that a theory can be disproven by new facts, but never proven. Absolute truth is beyond our reach. This is not so because we are too stupid to find positive proofs, but because of the structure of logic itself.

The "true believers" of the Velikovsky cult have never grasped this simple fact about science. To them, Velikovsky's correct predictions that Venus is hot or that Jupiter has a magnetic field have demonstrated the entire truth of his ideas beyond any possibility of a doubt. To them, the many other failures and contradictions of his theory do not matter, for they have never understood that the essence of scientific testing is falsification and not proof.

* * *

Now that distinguished scientists like George Wetherill are coming to believe that worlds have collided, as we described in the previous chapter, should we reassess Velikovsky's contribution? Should we acknowledge that, while he was wrong in detail, he was right in spirit? Did his publication of *Worlds in Collision* help set the stage for the new catastrophism? To address these questions, we must look specifically at the similarities and differences between Velikovsky's catastrophes and those now under study. And we must ask whether his ideas, his methodology, or his results actually affected—directly or indirectly—the scientists who have subsequently become more open-minded about the real role of catastrophes and chaos in the solar system.

The simple answer about any relationship between Velikovsky's catastrophes, which supposedly took place during the last several thousand years, and George Wetherill's colliding worlds 4.5 billion years ago is that there is a drastic difference in time scale: a factor of a million! (A factor of one million in time is the difference between a human lifetime, 85 years, and a single 45-minute class period in school.) There have been some impressive natural events in recorded history, like the explosions of the volcanoes Thera in 1645 B.C. and Krakatoa in 1883, or the Tunguska impact in 1908. But even such an event as the formation of Meteor Crater has not been witnessed in recorded history, and the asteroid fragment 200 feet across responsible for Meteor Crater was no planet.

Wetherill's colliding worlds represent a transient early phase of solar-system history, before the planets had accumulated most of the stray material. At that time, more than 4 billion years ago, the collision rate was extraordinarily high compared to most of planetary history. And, even then, there were millions of years between major collisions, while Velikovsky would have worlds interacting every 40 years or so.

What about Velikovsky's influence? It is difficult to be sure, for the human mind works in mysterious ways, but it seems more likely to us that Velikovsky's work inhibited an open-minded appraisal of catastrophism rather than assisted it. The astronomers who attacked the original book, and others who debated with Velikovsky's followers during the 1960s and 70s, were repelled by the obvious illogic and absurd conclusions of recent cosmic catastrophes. If anything, those scientists who felt it necessary to defend the status quo may have been dissuaded from considering any concepts of calamity. Most scientists, of course, simply ignored Velikovsky's proposals as crackpot ideas, like mental spoon-bending, horoscopes, and other products of popular purveyors of pseudoscience.

Velikovsky's approach was very nearly the opposite of that of the new planetary catastrophists. Believing unswervingly in his facile reordering of ancient history, Velikovsky readily admitted that his inferred astronomy required new forces and new laws, the nature of which he did not understand. George Wetherill, however, takes entirely the opposite point of view. As discussed in our chapter on chaos, Wetherill at first vigorously argued with his fellow scientists who proposed, without convincing reasons, that meteorites were derived from the asteroid belt. He argued, that is, until the rigorous mathematical calculations had been done that actually demonstrated how such transfers were possible. Wetherill, like most physical scientists but unlike Velikovsky, feels a compelling need to grasp the physical principles required to render a new idea plausible. In short, Velikovsky eschewed detailed analysis of the astronomical processes he invoked, whereas Wetherill insists upon it. Logical methodology is essential for supporting our modern views of catastrophism.

CHAPTER 14

Other Fringe Catastrophism

The theories of Immanuel Velikovsky appeal to our fundamental desire to humanize the cosmos. It is more than a little disconcerting to realize that the Earth is 4.5 billion years old and that humans have walked on our planet for less than 0.1% of this time. The historical record of humanity is, of course, even shorter, spanning only a few thousand years, or about 0.0001% of the age of the Earth. The great antiquity of our planet, and the relatively brief role played by humans in the grand cosmic drama, are disturbing and even threatening to many of us. In contrast, Velikovsky struck a popular chord when he suggested that events of truly cosmic magnitude have taken place within the few thousand years of human history, and that we can decipher these events from the historical records and oral traditions of our forebears.

Other fringe philosophers have also objected to the vast and impersonal cosmologies emerging from 20th-century science, and they too have proposed alternative interpretations, mostly catastrophic in content. Any compression of the geological time scale to human dimensions, which has been their goal, demands catastrophic mechanisms to produce the landforms and other visible characteristics of the world around us.

Historically, most religions also incorporated cosmologies that ascribed the origin of both the Earth and mankind to recent acts of divine creation. Among the major religions, only Hinduism and some versions of its great derivative, Buddhism, reject the idea of a creation and postulate instead a universe of indefinite age, characterized by long cycles of destruction and regeneration. In particular, the Judaic/Christian/Islamic religions of the Western world, as well as their apparent antecedents in the ancient civilizations of Sumer and Babylon, describe an origin of the Earth and humanity that apparently took place just a few thousand years ago, accompanied by catastrophic events of world-shaking proportions. These beliefs were a major source of inspiration to Velikovsky, and they were taken literally by most adherents of Judaism, Christianity, and Islam up until the 17th and 18th centuries, when the growth of science and the emergence of a more secular society led to reinterpretation of these concepts.

In general, 20th-century science and religion have reached an accommodation, in which each operates within its own sphere. Science teaches us about the universe in which we live and allows us to understand the processes that influence it, while religion deals with morals, ethics, and ultimate causes—areas outside the realm of science. This distinction is acceptable to most people, representing a wide variety of religious convictions. But during the past quarter century a curious reaction has developed, directed against both modern religion and modern science, based on a return to biblical literalism. This fundamentalist movement, which is largely confined to the United States, challenges our scientific worldview in fields as diverse as physics, astronomy, geology, biology, and medicine. Let us look briefly at this new catastrophist cosmology, based on the literal interpretation of the Old Testament.

* * *

Why should we discuss the opinions of a minority sect like the fundamentalists in a book on science? There are two reasons. First, many proponents of biblical literalism ask to be judged on the merits of their *scientific* ideas, even though these ideas are derived from divine revelation. They are convinced that the nature of the natural world, as explored by science, demonstrates that their ideas are correct and that conventional science, as it has evolved over the past century or so, is false. Second, they have coined a new term, "creation science," to describe their ideas, and they are widely promoting the inclusion of this "creation science" in public school curricula alongside conventional science, and with equal weight. Their political campaign has already had a major impact on American education, confusing teachers and students and contributing to a general "dumbing down" of American science textbooks, whose publishers are afraid to include material that might be controversial. So let us look at the ideas of "creation science" and their relationship to the other catastrophist concepts explored in this book.

One of the most popular expositions of the biblical literalists is to be found in *The Genesis Flood*, written in 1961 by fundamentalist theologian John C. Whitcomb and hydraulic engineer Henry M. Morris, who is now President of the Institute of Creation Research in San Diego. *The Genesis Flood*,* which has gone through more than 30 printings, presents a detailed case for a young Earth in which most of the geological and biological aspects of our planet were generated by a universal Flood (or Deluge) that took place about 6000 years ago. The starting point of Whitcomb and Morris, and of other creationists who have followed, is the Book of Genesis. They assert that science alone can tell us nothing of the past: "The only way man can have certain knowledge of the nature of events on earth prior to the time of the beginning of human historical records, is by means of divine revelation" (p. 439). And: "In the last analysis, the only really reliable recorder of time is man himself!" (p. 393).

Here is a summary of the beginnings of the world as presented by the creationists: Sometime less than 10,000 years ago (probably between 7000 and 8000 years according to biblical chronologies) God created everything—universe, Earth, life, and man—in six days. The Earth was perfect, a "Garden of Eden," and much more heavily populated than now. Its surface was relatively flat, and there were no oceans or ocean basins. The climate everywhere was moist, warm, and tropical due to a huge terrestrial atmosphere consisting primarily of water vapor and carbon dioxide, with a surface pressure many times greater than that today. This atmosphere is often called a "vapor canopy," a topic we will return to later. The vapor canopy contained that part of the primeval water that had, on the second day of creation, been collected into "the waters that are above the firmament." Some other quantity of water, perhaps much greater in extent, was separately trapped beneath the crust as "the waters of the great deep." This world was without rain, or winds, or storms of any kind. Its population contained at least a hundred times as many species as exist today, enough to account for the entire fossil record. Human lifetimes were measured in centuries, and the human population grew so fast that it spread around the globe within a couple of thousand years after the creation of the first couple, Adam and Eve.

According to the creationists, this world was whole and perfect, and curiously it included the illusion of great age—about 4 billion

*J.C. Whitcomb and H.M. Morris, *The Genesis Flood*, The Presbyterian and Reformed Publishing Company (1961), 27th printing.

years. This was supposed to be the age indicated by radioactivity in the rocks, for example. The surrounding universe simulated even greater age, with light arriving from stars and galaxies millions and even billions of light years away. In fact, the expansion of the universe was created to mimic an age of about 15 billion years. It is not clear why God chose to imbue His creation with these indications of vast size and great age, but some have suggested that it was a ploy to test our faith in the literal truth of the Genesis story. This concept, that the Earth and the universe present the illusion of age, is not described in the Bible, of course, but is an effort by the modern creationists to accommodate the discoveries of contemporary astronomy and geology.

Between 5000 and 6000 years ago, the story goes, God became dissatisfied with His creation. Having warned Noah to construct the Ark and bring together all of the species that were to be saved, he initiated the Deluge—the greatest catastrophe in history, which destroyed nearly all the life on our planet (a mass extinction) and gave rise to most of the phenomena studied by Earth scientists today, from sedimentary rocks to high mountains, from volcanoes to fossils, from the ocean and atmosphere to our concepts of weather and climate. The agent of all this destruction was the release of the waters that had been above and below the firmament, together with prodigious worldwide volcanism. "The 'fountains of the great deep' emitted quantities of juvenile water and magmatic materials, and the 'waters above the firmament,' probably an extensive thermal atmospheric blanket of water vapor, condensed and precipitated torrential rains for a period of forty days" (p. 328). Remember that this was the first rain that had ever fallen on the Earth! The water that was released, which now occupies the ocean basins, was sufficient (in the absence of high mountains) to cover the entire planet and to destroy all land plants and animals except those preserved in the Ark. Erosion from the Flood created all the Earth's sedimentary rocks within less than a year, and it formed all known fossils by burial of the creatures drowned in this catastrophe. Also formed at the same time were subterranean deposits of coal and oil.

After a few months the dry land reappeared, beginning with the tops of mountains. The Bible says that the waters receded, but since there seems to have been nowhere for them to go (presumably the vapor canopy could not have been reconstructed), modern creationists conclude that what really happened was a rapid elevation of the present continents. The scale of this uplift is remarkable, amounting to several hundred feet per day: "The termination of the Deluge proper, occupying a period of a little more than a year . . ., did not by any means mark the termination of the abnormal hydrologic and geomorphic phenomena . . . . The prediluvian topography was completely changed with great mountain chains and deep basins now replacing the formerly gentle and more nearly uniform topography" (p. 287). "Undoubtedly effects of these profound changes in the earth's surface and atmosphere were felt for centuries and are perhaps still being felt in some degree" (p. 270). The most important aftermath was a great ice age that came and went within a few decades, covering much of the northern hemisphere with ice and creating the extensive glacial deposits of our planet. All of this settled down, however, and by the time of the first written records from Sumer and Egypt (about 3000 B.C.) the world had assumed its present form and had been largely repopulated by the survivors of the Ark.

Again we turn to Whitcomb and Morris for an eloquent summary: "The picture is one of awesome proportions. The vast 'waters above the firmament' poured forth through what are graphically represented in the Scriptures as 'the floodgates of heaven,' swelling the rivers and waterways and initiating the erosion and transportation of vast inland sediments. At the same time, waters and probably magmas were bursting up through the fractured fountains of the great subterranean deep" (p. 265). "The Noachian Deluge was a cataclysm of absolutely enormous scope and potency and must have accomplished an enormous amount of geologic work during the year in which it prevailed over the earth" (p. 258). "A very substantial portion of the earth's structural geology must be explained in terms of the Flood, if the Bible record be true" (p. 270). "The Creation, the Fall, and the Flood constitute the truly basic facts, to which all of the other details of historical data must be referred" (p. 327).

* * *

If the arguments of the creationists rested here, as a statement of faith alone, we could dispense with this discussion. But in the name of "creation science," people such as Henry Morris and Wayne Gish (Vice President of the Institute for Creation Research and a prolific lecturer and public debater) insist that the Flood can be demonstrated by scientific evidence, and that their interpretation of the past is superior to that of conventional scientists. Since such claims are now being thrust upon legislatures and school districts, they warrant examination here.

A good place to begin is with the very young age—no more than 10,000 years—ascribed to the Creation, and even more recent dates for the Deluge, the formation of the present continents and mountains, and the global ice age. Can these be reconciled with other evidence? Of course not. We can begin with the Bristlecone pines that go back to 3500 B.C., and the Greenland ice cores that span more than 10,000 years, as described in the last chapter. But these are small problems compared with the weight of evidence for successive, and much older, ages of rocks in the geological column. These ages are determined, like those of Moon rocks and meteorites, from the relative quantities of radioactive elements and their decay products. Igneous rocks can be dated with good precision, with examples ranging from Hawaiian lavas erupted last week to the oldest continental rocks that go back nearly 4 billion years. Within the geological column, the radioactive dates confirm the sequence deduced from superposition of layers and provide an absolute time scale for organic evolution, the evidence of which is recorded by fossils in the sediments. (Creationists often claim that the dating of rocks involves circular reasoning, based on fossils that are assumed to represent an evolutionary sequence, but this is not the case. Relative ages were assigned by European scientists well before the concept of biological evolution was developed, primarily on the basis of the observed ordering of strata, while all absolute ages are calculated from radioactivity.) None of what we now know about either relative or absolute dates can be squared with the creationist model.

How do creationists answer such criticisms? There are two ways. The first and simplest is to deny the significance of the geochronometers. They suggest that radioactive decay rates may change with time, or that the measurements were done poorly, or that the technique is meaningless for any dates before the Deluge, when the world was fundamentally different. A more sophisticated alternative is to assert that the ages of the oldest antediluvian rocks are those emplaced by God when He made the created world appear old. All of the other, younger dates that span the geological column from 600 million years ago to the present represent errors introduced by the violent mixing of materials during the Deluge. To the scientists involved in measuring the ages of rocks, both of these attempts to dismiss all direct evidence for the great antiquity of the Earth seem ridiculous.

The idea that most of the Earth's sedimentary and volcanic deposits were produced during the short interval of the Deluge is another major problem for the Genesis Flood. The transformation of

Old and young rocks. At right is the Allende meteorite, which solidified 4.5 billion years ago; at left is a piece of Hawaiian lava that was erupted in 1988. So-called "creation science" denies the existence of rocks more than 10,000 years old and disputes all scientific methods of dating rocks and geological features. (David Morrison)

mud and sand into rock is a slow process, yet this single year is said to have generated literally miles of superimposed sedimentary layers of varied texture and composition. The Grand Canyon of the Colorado represents just a small portion of the geological column, yet who can hike into this mile-deep gorge and believe that all this was formed in a year (to say nothing of the carving of the canyon itself, which according to Flood chronology can only have begun in postdiluvian time, after the great uplift of the continents)? In the case of volcanic deposits, we know the cooling rates of lava from direct observations: It requires years to decades for a single layer a few feet thick to solidify near the surface, and it takes centuries to millennia for more deeply buried flows to cool. Yet the creationists insist that the thousand-foot-deep layers, composed of multiple lava flows, that make up such huge formations as the Snake River basalts in the northwest United States or the Deccan lavas of India formed in just a few months.

Another major stumbling block concerns the fossil record preserved in the sedimentary deposits laid down over the past billion years. According to the creationists, all of these plants and animals lived simultaneously and were killed and fossilized together during the Deluge, as the Earth's sedimentary rocks were formed. How can the creationist model explain the observed sequences of fossil types, as new (and often more complex) species are found in progressively higher (and younger) deposits? Here, as in the case of the measured ages of rocks, the most common recourse is to deny the existence of a fossil sequence. Critics of evolution like to find discrepancies in the record, especially those where subsequent mountain-building has folded and sometimes inverted the rock sequence. Having located these ambiguous examples, they reject the whole body of information from paleontology. The other approach is due to Henry Morris. Drawing upon his training as a hydraulic engineer, Morris proposes that the floodwaters sorted the drowned creatures so that the smaller ones sank more rapidly to the bottom and the larger ones were concentrated nearer the top of the forming sediment, thus mimicking in a general way the gross ordering of fossils. He suggests that the birds and mammals are near the top of the geological column because they fled to high elevations in the face of the onrushing waters and thus perished last. This speculation is ludicrous in light of the detailed studies of evolutionary progressions measured in the fossil records by scientists such as Raup and Sepkoski. The hydraulic sorting idea

The mile-deep Grand Canyon of the Colorado River. According to "creation science," the sedimentary rocks of the Colorado Plateau were all formed within a single year, and the entire Grand Canyon has been formed by erosion just within the past few thousand years. (David Morrison)

would be funny if it were not being seriously touted as science—and urged upon countless education boards across the United States as a part of the high-school science curriculum.

There are many other areas we could comment on, such as the problem of the dispersion of the animals from the Ark after it came to rest on Mt. Ararat in Turkey (how did the kangaroos get to Australia and the sloths to South America?), the impossible time scales for the raising of the mountains and the continents, or the compression of thousands of years of glaciation into a few decades. We wonder how the Earth could have supported a huge vapor canopy, and how the early humans managed to survive in a hot, humid atmosphere with many times its present surface pressure. Nor have we even touched upon the point that is central to much of this discussion: the evidence for biological evolution and the formation of new species. Most discussion of creationism focuses on evolution, a topic beyond the purview of this book. But, biological evolution aside, the geological case against a young Earth and a massive reworking of the planet by the Flood is overwhelming.

Of course, it helps if you begin an investigation or debate with an absolute assurance that you are correct. With such a conviction, you can select data that seem to support your case and reject all that do not agree. Such an attitude generates a remarkable level of self-confidence, as expressed again by Whitcomb and Morris: "The evidence of the reality of these great events, the Creation and the Deluge, is so powerful and clear that it is only 'willful ignorance' which is blind to it, according to Scripture. Thus do the Creation . . . and the Genesis Flood as indelibly recorded in human histories and in the rocks of the earth constitute the paramount scientific negation of all man-centered philosophy and religion for those who will accept it for what it is" (p. 453).

The new "creation science" as propounded by Morris, Gish, and their colleagues at the Creation Research Institute is different from the geological catastrophism of two centuries ago, against which Hutton and Lyell argued. Catastrophists of the late 18th century did not insist on a literal interpretation of the Genesis story, and they accepted the increasing evidence of the antiquity of the Earth. The great French anatomist George Cuvier, for example, suggested that the Earth's history had been punctuated by a series of catastrophes, each of which resulted in the widespread destruction of life and the emplacement of another layer of sediment in the geological column. The most recent of these catastrophes was thought to be the biblical Deluge. The new scientific perspective we have been discussing, involving asteroidal impacts, mass extinctions, and punctuated equilibrium in the evolution of life, bears some resemblance to these ideas, although it certainly differs in many other respects. But neither this new perspective nor the traditional catastrophism of 200 years ago has anything to do with the pseudoscience peddled under the rubric of "creation science." Indeed, we agree with Steven Jay Gould, who has labeled "creation science" an oxymoron—a self-contradictory phrase, such as "good grief," "postal service," or "limited nuclear war."

* * *

Creation science does not represent the only catastrophist cult of recent times. One of its antecedents, for example, lies in the work of the 19th-century American Quaker Isaac Newton Vail. In his 1886 book *The Waters Above the Firmament*, Vail formulated the idea that the Earth originally had a ring of water or ice rather like that of Saturn. The collapse of this ring provided the waters for the Deluge. Vail's ideas are still popularized by the Annular World Association of Azusa, California. Even more influential on creationist thinking was the geology text published in 1923 by George McCready Price, a professor at a Seventh Day Adventist college in Nebraska. In *The New Geology*, Price developed many of the ideas later popularized by Whitcomb and Morris, such as the misconception that all geological dating represents a process of circular reasoning based on a presupposition of biological evolution. In a sense, Price's book is not *about* geology but *against* it, since he devotes lengthy chapters to countering the accepted wisdom in this field. In the end, Price concludes that the Deluge was itself responsible for essentially all of the geological features of the Earth today.

One of the more disturbing episodes in fringe catastrophism unfolded in Germany in the 1930s. This particular piece of pseudoscience goes by the name of WEL, the acronym for Welt-Eis-Lehre (Cosmic Ice Theory). Its author was a Viennese mining engineer, Hans Horbiger, who published his 790-page opus *Glazial-Kosmology* (Ice Cosmology) in 1921. The theory centers on the idea that the Earth has had a series of icy natural satellites, each of which eventually impacted our planet, with catastrophic results. These bodies are imagined to have originated in the outer solar system, which accounts for their icy composition. At any time, there is a whole string of them spiraling inward toward the Sun, due to a supposed drag from interplanetary hydrogen gas, and on the way some of them are captured into Earth orbit. Others, which reach the Sun itself, somehow cause the sunspots. Once captured, these icy moons spiral in toward the Earth until they are destroyed. According to Horbiger, the previous moon (before our present Luna) met its end in late prehistoric times, when it gave rise to various legends of dragons and celestial battles. Its final demise precipitated the biblical Deluge, in which nearly all life was destroyed. There followed a period of calm, according to Horbiger, associated with the myths of paradise and Atlantis. About 13,500 years ago our present Luna was captured, precipitating a new series of global catastrophes that marked the beginning of our present epoch of history.

One strange aspect of this particular aberration was its great popularity during the Nazi period in Germany. Followers of WEL published books and pamphlets, held rallies, participated in politics, and agitated against academic ("Jewish") science. After World War II this theory was less actively promoted, but there may still have been millions of WEL believers who looked forward to the discovery, from the Apollo Program, that the Moon might be composed of ice. But of course, Apollo demonstrated that the Moon is not only ice-free but also one of the most desiccated objects in the solar system. Whatever the origin of our Moon, it is not a captured icy body from the outer solar system.

Although we like to think that people are more scientifically literate today, they are not immune to the appeals of pseudoscientific catastrophe theories. A case in point is the worldwide concern over the "great planetary alignment" of the early 1980s. In a book called *The Jupiter Effect*, published in 1974, science writers John Gribbin and Stephen Plageman suggested that a special configuration of the planets to take place in 1982 would likely trigger catastrophic earthquakes, a prediction that was taken particularly seriously in southern California. The theory was that the planetary alignment would cause unusual tides on the Sun, disturbing the solar activity cycle and increasing the numbers of sunspots and solar flares. Such solar events would heat the Earth's upper atmosphere, which would expand, thereby altering slightly the Earth's rotation. Although tiny, such rotation changes would set up internal forces, triggering earthquakes in areas under high stress, such as along the San Andreas fault in California, or so Gribbin and Plageman wrote.

The so-called "planetary alignment" consisted of the fact that all nine planets were approximately on the same side of the Sun for a few days in early March 1982. Actually, the planets were stretched over an arc of 95°, or more than one fourth of a complete circle. It is easy to show that, even if the alignment were better, the tidal effects on the Sun would be negligible—less than 0.1 inch, in fact. Nor is there any evidence that tides influence solar activity, or that changes in the Earth's atmosphere trigger earthquakes. The entire logic is no more than a house of cards, completely without merit. When critics pointed out the fallacy of these arguments, author Gribbin conceded that there was only a slight probability of increased seismic activity in 1982. However, *The Jupiter Effect* was not withdrawn from publication, and when March 1982 rolled around, millions of people expressed their interest and, often, concern.

As the millennial year of 2000 approaches, we can expect an increase in the predictions of imminent catastrophe. There is something in such concepts that appeals to people. Perhaps it is the same urge that attracts us to horror films: a desire to be frightened. The universe is indeed a big and powerful place, and there are real dangers to consider, as described throughout this book. Let us retreat from our detour into pseudoscience and return from the realm of fantasy to the universe of reality.

CHAPTER 15

Climates Gone Wrong: Venus and Mars

Among the planets of our solar system, the Earth is unique in many ways, including the presence of oceans of liquid water, an oxygen-rich atmosphere, and abundant life. But there are two other planets that resemble our own in many other ways: Mars and Venus. A prime motivation for exploring the solar system has been a desire to improve our understanding of the Earth by comparison with our planetary neighbors. At the beginning of the space age, many scientists anticipated that Venus and Mars would turn out to be basically similar to our own planet. One of the most important lessons of nearly 30 years of planetary exploration has been the realization that this expectation was wrong.

The atmospheres and surfaces of both Venus and Mars are extremely inhospitable, one an inferno of oven-hot temperatures and poison acid clouds, the other bathed in lethal ultraviolet radiation and frozen in a perpetual ice age. If we are to protect and preserve our own atmosphere and climate from catastrophe, either natural or human-made, it is vital that we understand how our two most Earth-like neighbors went so terribly wrong in their atmospheric evolution.

Each of these two planets has been visited by more than a dozen spacecraft, first by simple flybys, then orbiters, atmospheric probes, and eventually landers that operated on their surfaces. Both of the major spacefaring nations, the United States and the U.S.S.R., have made important contributions to this effort; the Americans have concentrated their missions toward Mars, while the Russians sent 18 of their Venera spacecraft to Venus before turning to Mars in the 1990s.

Let us compare some of the basic properties of Mars and Venus. Mars orbits the Sun at an average distance of 142 million miles, about

50% farther out than the Earth. As a planet, it is rather small, with only 11% of the mass of the Earth. It is substantially larger than the Moon or the planets Mercury and Pluto, however, and is large enough to retain a thin atmosphere. As we have seen, Mars is also large enough to have generated the internal heat necessary to support volcanism and other forms of geological activity, some of which apparently persist to the present day. This planet, of course, has historically been the subject of much speculation concerning indigenous life, especially in the decades around the turn of the century, when the enthusiasm of that proper Bostonian Percival Lowell led to wide interest in a "canal"-building race of intelligent Martians. Lowell envisioned Mars as having a climate that was generally Earth-like, albeit somewhat drier and colder, like the high desert plateaus of Tibet or the dry valleys in Antarctica. Although his fantasies concerning canals and a martian civilization scarcely outlived his death in 1916, the idea that Mars might support some kind of life persisted until the first Viking landed on the surface in 1976.

Since Venus orbits almost twice as close to the Sun as the Earth, one might expect it to be warmer than our planet. However, its bright clouds reflect away so much sunlight, one could alternatively imagine it to be cooler. Venus is much larger than Mars, and just a tiny bit smaller than our planet. Geological processes are still active, as revealed by recent studies of its surface carried out by the Soviet Venera landers and radar-mapping orbiters. Only in the 1960s did new data demonstrate that the surface temperature on Venus is far above the boiling point of water (nearly 1000° F, in fact), and that the clouds are composed of deadly sulfuric acid. Venus's cloud-shrouded atmosphere of carbon dioxide is vastly more massive than ours; the surface pressure is 90 times the sea-level pressure on Earth.

Up to the 1950s, however, Venus was often pictured as quite Earth-like, lush and humid like our tropics. Science-fiction writers such as Edgar Rice Burroughs popularized the notion of dense jungles populated by all manner of exotic creatures. Implicit in the earlier popular conceptions of Venus and Mars were two assumptions. First was the idea that nearby planets would have similar surface conditions to the Earth's, a little hotter or colder or dryer or wetter, perhaps, but generally within the range of climatic conditions we encounter in diverse parts of our own planet. The second implicit assumption was that life (usually pictured as rather like that of the

Earth) was also a common phenomenon, to be expected on planets where conditions were suitable. Neither of these comforting myths has survived scrutiny in this age of direct space exploration of our neighbor worlds.

* * *

Venus has often been called Earth's twin. Certainly the two are alike in their size, mass, and bulk chemical composition. Theories about the origin of the solar system suggest that the two planets must have begun with similar surface conditions as well: silicate rocks, widespread volcanism, oceans of liquid water, and thick atmospheres composed mostly of carbon dioxide. Yet somewhere their evolutionary paths diverged, as we learned when the first measurements of the surface temperature of Venus were made by radio astronomers in the late 1950s. Instead of the expected temperatures of 100–200° F, the radio emission from the planet indicated values of more than 800° F. In the early 1960s this extraordinary heat was confirmed, first by the Mariner 2 flyby, and later by the Soviet Venera entry probes and landers. The surface of Venus is hot enough to melt lead and zinc. Other evidence revealed that the thick clouds that obscure the atmosphere are composed, not of water droplets as many had thought, but of sulfuric acid.

Rather than being a tropical paradise, Venus resembles a vision of hell. Everything is shrouded under a dense reddish overcast; the atmosphere is heavy, unbreathable, sulfurous; the surface is nearly hot enough to glow by its own heat. The several Soviet spacecraft that successfully landed there directly measured these inhospitable conditions during the few hours available to them before they succumbed to the terrible heat and pressure. Although a few people, Immanuel Velikovsky among them, speculated that the surface was being heated from below in some global volcanic outburst, most planetary scientists recognized from the start that the high temperature on Venus must be the result of ordinary sunlight, trapped near the surface by the thick atmosphere. This conclusion is supported by the measured absence of any net loss of heat from Venus: The clouds and upper atmosphere are no warmer than we would expect for a planet at Venus's distance from the Sun. Instead, the heat is confined to the surface and the dense lower atmosphere. In the 1960s, planetary scientists like Carl Sagan and James Pollack (both later involved in the nuclear winter debate) set out to calculate how the atmospheric blanketing worked.

Imagine the fate of a sunbeam that diffuses through the atmosphere of Venus and strikes the surface. It is absorbed, warms the surface, and is re-emitted as heat in the infrared part of the spectrum. However, carbon dioxide, which is a colorless, transparent, and invisible gas to our eyes, is opaque to infrared rays. As a result, it acts as a blanket, making it very difficult for the infrared radiation to leak back to space. The surface retains the energy until its temperature becomes so high, and it radiates so much heat, that the little fraction that escapes balances the incoming sunlight. This process is often called the "greenhouse effect," because it resembles the effect used for centuries to warm a gardener's greenhouse and extend the effective growing season in cool climates. Many of us are most familiar with the greenhouse effect because it heats the inside of a car left out in the Sun with the windows rolled up. In these examples, the window glass plays the role of carbon dioxide, letting sunlight in but impeding the outward flow of infrared radiation. We all know one result: a hot seat when we sit down in a car interior that is much hotter than would otherwise be expected from solar heating.

The atmospheric greenhouse effect adequately explains Venus's high surface temperature, and it is amply confirmed by both direct measurements and detailed computer models. But it begs the more difficult question of how Venus reached this state in the first place. Scientists have reason to believe that the atmospheres of Venus and the Earth were initially similar, and the records of ancient climates on the Earth also suggest that our own planet once had more carbon dioxide and a larger greenhouse effect than exists today. In fact, it appears that the declining terrestrial greenhouse effect during the past several billion years has nicely compensated for a slow increase in the luminosity of the Sun over the same period, maintaining a nearly constant surface temperature on our planet. Clearly, no such balance was achieved on Venus, and we naturally ask ourselves if there is any danger of the Earth evolving into a Venus-like condition, which would certainly be lethal to all life on our planet.

* * *

One way scientists approach the question of the differences between Venus and the Earth is to investigate the possible consequences of changing the Earth's present atmosphere, or the early atmospheres of Earth and Venus. Both planets probably began with

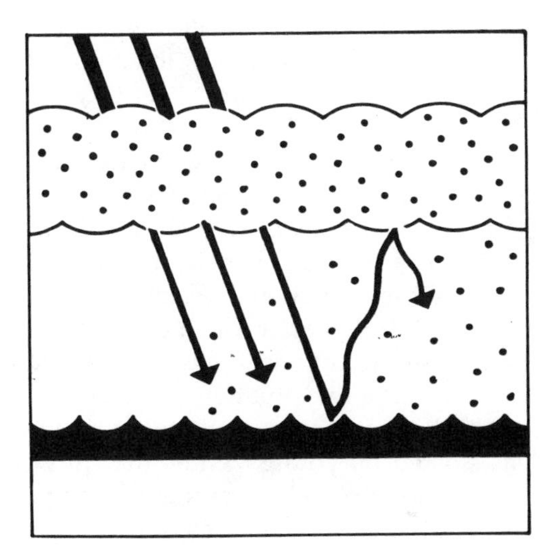

The atmosphere transmits the sunlight, but tends to hold in the reradiated heat waves, thus warming the Earth. (David Fischer)

moderate surface temperatures and extensive seas of liquid water, before their evolution diverged. How easy is it to get from one state to another; how short the path from paradise to hell?

Imagine that the Earth or an Earth-like Venus is heated, for example, by a small rise in the energy output of the Sun. One consequence is a further increase in atmospheric carbon dioxide and water vapor, as a result of increased evaporation from the oceans and release of gas from surface rocks. These two gases would in turn produce a stronger greenhouse effect, further raising the temperature and leading to still more CO_2 and H_2O in the atmosphere. Instead of stabilizing or correcting the original small change (an increase in temperature, in this example), Nature reacts to accentuate or amplify it. This is called "positive feedback," and it is generally a bad thing, because it leads to instability.

The situation where a small increase in the greenhouse effect is amplified rather than corrected has been aptly termed a "runaway greenhouse effect." The runaway greenhouse is not just a larger greenhouse effect; it is a process whereby an atmosphere shifts from an initial state where the greenhouse effect is small, such as on the Earth today, to a much hotter state, such as we see on Venus. Once the larger greenhouse conditions develop, the planet establishes equilibrium, and reversing the situation is difficult if not impossible. Many scientists think that this is exactly what happened on Venus.

If large seas existed on young Venus, the runaway greenhouse would have caused them to boil, accentuating the positive feedback from the CO_2 by adding large quantities of hot water vapor—which also contributes to the greenhouse—to the atmosphere. Water vapor is not stable in the presence of solar ultraviolet light, however, which breaks apart H_2O molecules into their constituent parts, oxygen and hydrogen. The light hydrogen escapes into space from the top of the atmosphere, leaving the oxygen to combine chemically with surface rock. The loss of water is, therefore, also an irreversible process; once the hydrogen has vanished into space, water cannot be restored.

Venus provides us with a cautionary tale illustrating an extreme of atmospheric evolution. One possible consequence of tinkering with the Earth's delicate balance is starkly before us, not as the result of some esoteric computer models, such as those used to predict a nuclear winter, but as a real planet, once similar to our own, but now with a suffocating atmosphere and a surface temperature of more than 800° F. When we look outward toward our other planetary neighbor, Mars, we see a similar cautionary example representing the opposite extreme: a planet that has lost most of its atmosphere, with its surface now eternally frozen.

* * *

When the Mariner 9 spacecraft went into orbit around Mars in 1969, the red planet was still largely an enigma. Previous flybys had provided glimpses of its frigid surface, pocked with impact craters, and had probed its thin, dry atmosphere of carbon dioxide. Most scientists had looked at the photos of a lunarlike surface and quickly abandoned the century-old hope that Mars might harbor life, except perhaps at the microbial level: Its atmosphere was too tenuous, its surface too dry and cold. But the Mariner 9 orbiter was designed to do much more than radio back a few snapshots. It undertook a two-year global survey of Mars, which revealed a much more complex and interesting planet. To be sure, Mars is cold and dry, and most of its surface is heavily cratered. But in addition there are huge volcanoes, globe-girdling canyon systems, great sand seas, and many other phenomena that fascinate planetary geologists. A particularly exciting and surprising discovery was ample evidence of ancient watercourses, now desiccated and silent. Whatever its state today, Mars was not always so inhospitable. Long ago, its atmosphere was denser and warmer, clouds formed, and rain fell. Great rivers meandered across the landscape, emptying into ancient seas. Perhaps life flourished, as it did in the archaic seas of the Earth. But Mars, like Venus, is a planet that failed, at least by our human standards. One of the great challenges of planetary science is to understand why.

In 1976, the United States sent four Viking spacecraft to Mars, two advanced orbiters and two landers. The orbiters confirmed and extended the Mariner discoveries, while the landers settled gently onto the plains named Chryse and Utopia for detailed looks at the ground, including a search for microbial life at these two sites. The results of the life-detection experiments were negative, but a great deal was learned about other aspects of the planet. Weather stations on each lander measured temperature, pressure, and wind. As expected, temperatures vary much more on Mars than on Earth, due to the absence of moderating oceans and clouds. Typically, the summer maximum was 10° F, dropping below −100° F just before dawn. The lowest air temperature was −140° F. Even worse from a biological perspective, the thin atmosphere lacked both oxygen and ozone, permitting unattenuated solar ultraviolet radiation to strike the surface, sterilizing the soil and eliminating any chance of indigenous life.

The atmosphere of Mars, which is composed of 95% CO_2, has a surface pressure less than 1% of that of the Earth. Although there is a trace of water vapor in the atmosphere and occasional ice clouds form, liquid water is not stable under present conditions on Mars. For one thing, the temperatures are too cold. But even on a rare sunny summer day, when temperatures rise above the freezing point, liquid water cannot exist. At the low atmospheric pressure, the boiling point of water is as low as or lower than the freezing point, so water sublimes (that is, it changes directly from solid to vapor without an intermediate liquid state).

Seen through a telescope, the most prominent surface features on Mars are its bright polar caps, which grow much larger in the winter, similar to the winter snow cover on Earth. We do not usually think of snowfields as part of our polar caps, but viewed from space, the thin snow would seem to extend the thick permanent ice caps, as seen on Mars. However, the seasonally varying caps on Mars are composed not of ordinary snow, but of frozen carbon dioxide (dry ice).

Distinct from the thin seasonal caps of dry ice are permanent, residual polar ice caps that are present even in summer. As the seasonal cap retreats during spring and early summer, it reveals a brighter, thicker cap beneath. The southern permanent cap is composed of thick deposits of frozen carbon dioxide, but the northern permanent cap is different. Much larger, it is composed of ordinary water-ice. We do not know its thickness, but it may be as much as several miles. In any case, this permanent cap represents a huge reservoir of water, in comparison with the very small amounts of water vapor in the atmosphere. Even more water is thought to be hidden below the surface in the form of permafrost (permanently frozen soil). Unlike Venus, Mars has not lost its water; it has simply locked it away in the inaccessible forms of polar ice and deep permafrost.

Yet in spite of these present conditions, the geological evidence of the Mariner and Viking photos tells us that once long ago rain fell and rivers flowed on Mars. These more clement conditions probably existed at least 3 billion years ago. On the geological time scale, Mars experienced a short and perhaps glorious spring, but summer never followed.

The ancient cooling of Mars and the loss of its atmosphere result from its small size relative to the Earth and its greater distance from the Sun. Mars formed with a much thicker atmosphere, which for a time maintained a higher surface temperature (as on Venus and Earth) due to the greenhouse effect. Partly due to its lesser gravity, its atmosphere soon escaped to space, which gradually lowered the temperature. Eventually it became so cold that the water froze, further reducing the atmosphere's capacity to retain heat. The result is tne cold, dry planet we see today. Unlike the runaway greenhouse effect of Venus, Mars experienced a sort of runaway refrigerator effect. From our human perspective, however, the results were just as intolerable.

* * *

What do we make of the fact that our two neighbor planets, which started out with atmospheres and climates not too different from Earth's, evolved to such different climatic extremes? Are we extremely lucky that our planet happens to be balanced in between, so that life could develop and flourish? One study several years ago concluded that it was simply good fortune that our planet happened to be formed within a very narrow zone of habitability. The research indicated that if the Earth had formed just a little bit closer to the Sun, there would have been a runaway greenhouse, just as happened on Venus. And if the Earth had been as little as 1% farther out, there would have been runaway glaciation, our oceans would have frozen solid, and our planet would have become as inhospitable as Mars.

These conclusions had profound implications for the prevalence of life in the universe. If other solar systems form and evolve in ways analogous to our own, there is the possibility of life evolving on planets throughout our galaxy and the universe. But if the zone of habitability, between broiling and frozen planets, is extremely narrow, the chances are that most planetary systems would lack a habitable planet. Only a few systems would have a planet at just the right distance to balance between the climate extremes and provide potential abodes for life.

More recent studies, by a scientific team that includes James Pollack and Brian Toon of TTAPS fame, suggest instead that the Earth's climate *may* be a self-correcting system, less prone to catastrophic runaways. Instead of positive feedback, these researchers find evidence for some negative feedback in Nature. The Earth is a very complex ecological system, however, and many scientists are much more fearful that perturbations by human civilization, in partic-

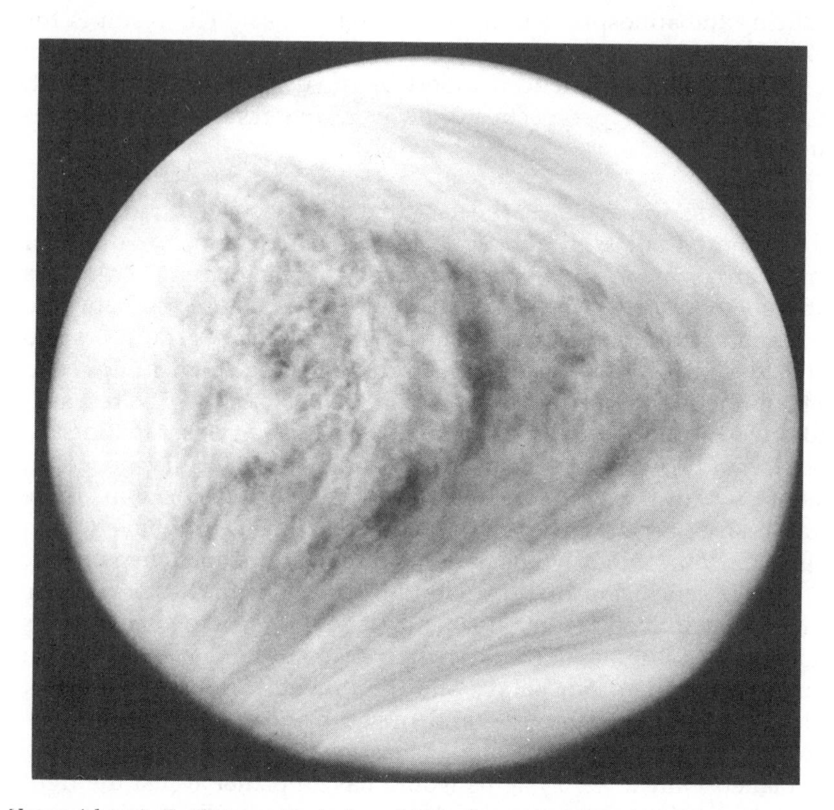

Venus (above), Earth (opposite top), and Mars (opposite bottom), shown to approximate scale. The thin atmosphere of Mars is revealed here only by the ice clouds associated with a large volcano, and by frosts and fogs at high latitudes. (NASA)

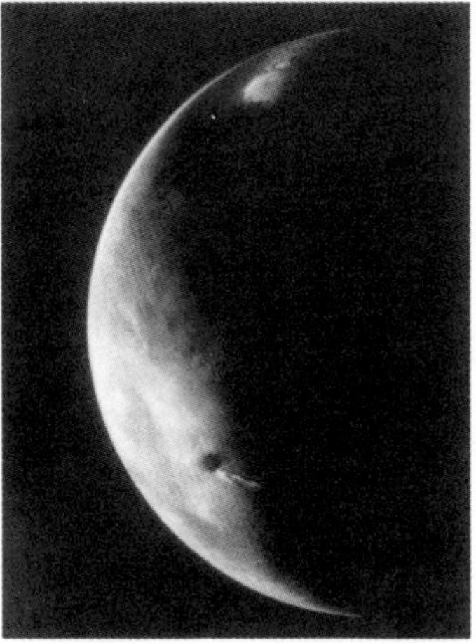

ular, could upset our applecart. It may be that the zone of habitability for planets is initially rather broadly encompassing. But a new factor has entered the equation, human beings. From studies of the response of our planet's oceans, weather, forests, and other components to natural oscillations of the past, like ice ages, as well as to the influences of modern civilization, we hope to learn about how robust, or fragile, our world is. We hope that scientific research will reveal the answers before we unwittingly stumble into a runaway change, like one of those that affected our planetary neighbors, Venus and Mars.

CHAPTER 16

The Greenhouse Effect and Ozone Holes

In early summer 1988, it was hot. In Ohio and Michigan, the temperatures climbed well into the 100s day after day. In the northern plains, the wheat crop was failing, dust was blowing, and fires were burning after months of drought. The nation's lifeline, the Mississippi, shrank in places to a shallow trickle in an expanse of drying mud. In Tucson, Arizona, the temperature hit 114° F; it had never been hotter than 112° F before, not just on that date, but ever. By August, the temperature of Lake Erie's waters rose to the warmest ever. Around the globe, the story was much the same.

Media pundits were near panic at the prospects of a new Midwest dustbowl, like that of the 1930s. Scientists testified to Congress about their predictions of broiling heat in the next century. Worst of all, the cause was attributed not to a rare fluke of nature but to the thus-far-uncontrolled environmental impact of our own civilization. The sweltering heat could at last be related to a planetary catastrophe in the making. *Newsweek* put it in cosmic perspective: "We forgot the one force capable of upsetting the balance that it took those billions of years to create. Ourselves."

Is this for real? We have all heard doom-and-gloom predictions before, which failed to materialize. What happened to the gas-station lines of the early 1970s? Black Monday's stock market crash of 1987 failed to usher in another Great Depression. Threats of war come and recede. Maybe 1988 was just the year of the "greenhouse fad" and the news of the '90s will be of blizzards and a new ice age. The weather, after all, is notoriously fickle and given to extreme oscillations. Every day, new weather records—for both hot and cold, wet and dry—are set somewhere in the world. Aren't we just as likely to be reading of floods along the Mississippi in years ahead as we are about more barges stranded on mud flats?

The answer, of course, is that the great greenhouse scare of 1988 was both faddish and factual. There will be floods again on the Mississippi and there will be frigid winters again in Iowa. Nevertheless, world temperatures are increasing, slowly but surely, and scientists are nearly certain that they know the reason: Chiefly it is because there is more carbon dioxide in our planet's atmosphere due to our burning of fossil fuels (oil and coal) and to the destruction of tropical forests. The trace gas carbon dioxide (CO_2) is produced when carbon left over from burned fuels combines with atmospheric oxygen. In a natural cycle, CO_2 is consumed by photosynthetic plants and, in the past, consumption by plants has roughly balanced natural production. Now, partly due to diminishing plants caused by deforestation, CO_2 has increased by nearly 50% in the last century. That is a change greater than the largest natural variations, which are a result of the periodic ice ages. At a third of a part per thousand, CO_2 is more abundant now than it has ever been in at least 160,000 years, and it may even be more than at any time since *Homo sapiens* first evolved.

CO_2 acts as an effective blanket, keeping the Sun's heat near the ground. Unlike our more common atmospheric gases, which let sunlight in and also let heat radiate back out to space, CO_2 lets the sunlight in but is opaque to heat radiation. The predicted warming has now been verified, beyond any statistical doubt that it was due to unlucky chance: Until stifling 1988, 1987 was the warmest year on average around the globe, since reliable temperature records have been kept. And all of the five hottest years of the past century have been in the 1980s. Not only are weather stations recording hotter temperatures, but another prediction of the greenhouse effect seems to be verified as well: Balloon measurements seem to show that the stratosphere, above the CO_2 blanket, is beginning to cool.

The climatic warming has been very gradual this century. There have even been some colder spells as recently as the 1950s. Ordinary variations of weather and climate will continue to produce fluctuations, but due to the influence of the global greenhouse, the average temperatures will climb inexorably. Within 20 years, world temperatures may be hotter than they have been for 100,000 years—unless we take concerted, worldwide action. By 2030, compared with the pres-

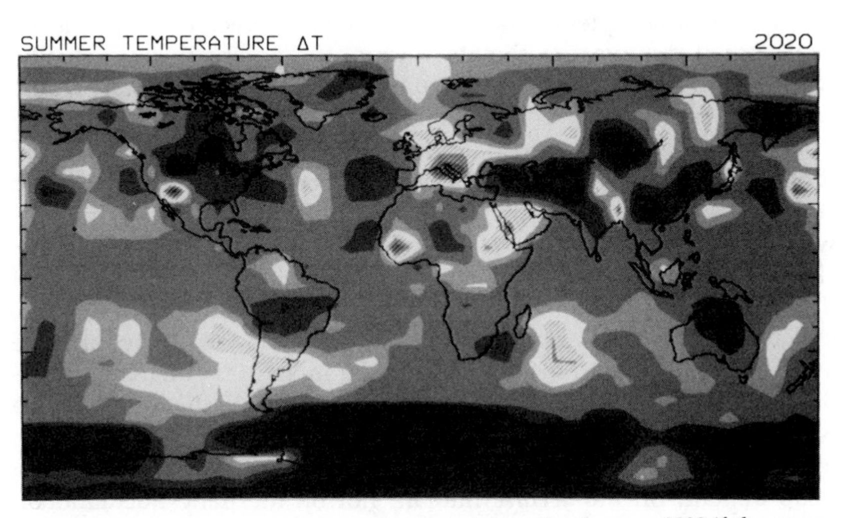

Calculations of increases in the summer temperatures by the year 2020 if the greenhouse effect is not controlled. White areas show little change. The darkest regions will be 5 to 10 degrees F warmer. Hatched regions will actually be colder by a few degrees. (Jim Hansen, Goddard Institute for Space Studies)

ent, New York City can expect three times as many days with highs in the 90s or 100s, but that is not to say there won't be occasional low-temperature records set even then. There will be hellish days, when an ordinary heat wave piled on top of the warmer climate will send thermometers soaring higher than ever before, with attendant heat strokes, power shortages, and so on. The larger effects on lives of our children and grandchildren, however, will be due to those aspects of our planet, and our economy, that are sensitive to temperature *averages*.

The enormous volumes of water in the oceans respond very slowly to changing temperatures from week to week or even year to year. But if the average climate warms, the oceans will eventually warm, too. Ocean waters will expand (as most materials do when warmed), arctic ice packs may melt, and the sea level will surely rise. Much of the world's population resides near sea level, so even though the predicted rise—nearly four feet by the middle of next century—sounds small, the homes of tens of millions of people would be inundated. Modern agriculture has been finely tuned to the climate, and will be especially vulnerable to warming. Scientists predict the breadbasket and cornbelt parts of the United States to have climates that cannot sustain the crops grown there today, primarily as a result of more rapid evaporation of soil moisture. Canada and the Soviet Union may benefit, but desertification of productive drylands will be the rule rather than the exception worldwide. Already 15 million acres of new desert are forming each year. Also, water reservoirs depend on the *average* rainfall, not on the daily fluctuations. Southern California and Arizona are among the growing sunbelt localities that rely on Colorado River water, but its watershed may have only 60% of the rainfall that it enjoyed when the costly water projects—like the Central Arizona Project canals—were planned and built.

Greenhouse heat is not the only impact of civilization on our planet's environment. There is also acid rain, which is killing trees throughout Europe and eastern North America. There is the poisoning of our lakes and aquifers, and the increasingly dirty air in our growing cities. Even outer space is being polluted; a chart of space debris shows our planet already enveloped by a cloud of rocket fragments and dead satellites, like a cow surrounded by a swarm of flies. As the world population climbs toward 6 billion, we need to know whether environmental protection is merely a conceit of the well-to-do, concerned about their "quality of life," or instead if we are truly participating in the final destruction of the planetary environment that has fostered evolution of life, and made our civilization possible. Certainly life will remain tolerable in many parts of the world for some decades to come, but could we be contributing to the eventual "venusification" of Earth? In our can-do, pro-growth era, many people seem to believe that technology can save us from whatever dangers are ahead. But others wonder whether we, as a species, are technically, economically, and politically capable of controlling our effects on the Earth, of changing our old habits, or have we passed the point of no return?

* * *

The Earth, Mars, and Venus formed from roughly the same cosmic mix of gas and dust, which made up the primordial solar nebula 4.5 billion years ago. Some of the gases that now reside in these planets' atmospheres were incorporated into their interiors when they were forming: The asteroidlike planetesimals from which planets were made contained gases. Afterward, icy comets and asteroids continued to rain down on the young planets, adding more volatiles to their surfaces. Oceans and atmospheres developed from this uncertain mix of original and cometary volatiles, some of the liquids and gases oozing from the planet's interiors during early eras of tumultuous volcanism. The atmospheres of all three planets were originally mostly CO_2, along with water vapor, carbon monoxide, and perhaps some hydrogen-containing gases like methane and ammonia. Solar ultraviolet light attacked the carbon monoxide, methane, and ammonia, leaving mostly immense quantities of CO_2 and water.

Then the histories of Earth, Mars, and Venus diverged. As we have seen, Mars originally had a clement climate still evident in the river valleys that have eroded its ancient cratered terrains. But it could not hold onto its greenhouse gases; its thin, tenuous CO_2 atmosphere has far too little water vapor or CO_2 to create greenhouse warming. The red planet froze long ago and remains a dry, frigid, lifeless world. On Venus the reverse took place, a runaway greenhouse that led to its oceans boiling away and a permanent loss of water. Broiling Venus is now enshrouded in its dense CO_2 atmosphere, containing only small percentages of nitrogen and argon.

The Earth's history has been very different. Only on our planet did the temperature stabilize within the narrow range that makes liquid water, and life, possible. As the Sun gradually brightened over the aeons, the growing population of marine organisms slowly but steadily extracted CO_2 from the atmosphere, thereby lessening the greenhouse effect. Eventually photosynthetic life, which feeds on CO_2, proliferated to the point that nearly all of the CO_2 was gone or buried, and oxygen gas began to take its place as a major component of the atmosphere. Oxygen in the upper atmosphere reacted with solar ultraviolet light to form the protective layer of ozone, which shields the surface of our planet from the Sun's deadly rays and made it possible for life to emerge from the oceans and spread out across the once barren, sterile lands of our planet. Thus on Earth, the presence first of water and then of life led to our unique atmosphere. Within limits, our planet's atmosphere has been fairly stable ever since, even stable enough for life to survive on land for hundreds of millions of years. Just *how stable* is our ecosystem? Could we be in danger of destabilizing it, to our descendants' doom?

Of course, the climate has not been constant, or entirely stable. During the Cretaceous period, prior to the great impact 65 million years ago that changed the history of life on our planet, dinosaurs reigned in tropical forests that thrived in now-temperate latitudes. Alligators lived near the Arctic Circle, and sea levels were a thousand feet higher than today. Part of the reason may have been an enormous, natural influx of CO_2 from the depths of the Earth; geological evidence suggests that volcanic activity was high at the time. Our planet's climate has tended to cool over the ensuing tens of millions of years. Beginning between one and two million years ago, cycles of ice ages began. The fluctuations of ice ages every 100,000 years or so indicate some tendency for Earth's climate to oscillate only within limits.

What sets these limits? Why doesn't our planet, when it is in the midst of an ice age, follow the course of Mars and become ever colder and drier? One might imagine that the widespread, brilliant snow would reflect away the sunlight, reinforcing the cold, resulting in more ice, and so on until the climate had spiraled into a permanent deep freeze. On the other hand, why doesn't a runaway greenhouse start at the peak of interglacials, when the temperatures are warm and CO_2 is abundant? The latter question is easier to answer than the former. During past interglacial periods, for which there are geological records, the temperatures never became as warm as they were during the Cretaceous. Since a runaway greenhouse didn't happen during the Cretaceous, evidently the interglacials never approached the as-yet-unexplored threshold for such a disaster. Alarmingly, human civilization already seems to be making the present interglacial the warmest on record, and we could be testing the limits of the Cretaceous within a couple of centuries, if present trends were to continue.

There have been about ten ice ages in the last million years. The most recent ice age ended, in a rather worrisome way (as we will see), just about 13,000 years ago. Scientists are not entirely sure about what causes the 100,000-year cycle. But about 60% of the fluctuations in the amount of land-ice can be attributed to well-known, and seemingly minor, astronomical causes. In the early decades of this century, a Yugoslavian astronomer, Milutin Milankovitch, proselytized his belief that the ice ages were due to small variations in solar energy reaching various latitudes on our planet, caused by natural oscillations in Earth's orbital motions and the tilt of its axis. The tilt of the axis, now 23.5°, varies regularly every 40,000 years between 22.1° and 24.5°. Every 26,000 years, the Earth's axis completes one circular "wobble" (like a wobbling top) in the direction it is pointing in the sky, which is now toward the star Polaris, which has been the closest "north star" for just a couple of thousand years. And on a cycle of about 100,000 years, the shape of the Earth's orbit varies from almost perfectly circular to slightly (6%) elongated. Milankovitch argued that these astronomical oscillations, small though they are, could have major climatic effects in the high latitudes where glaciers grow and retreat.

Consider this example. In northern Canada, it is cold and snowy in the winter, regardless of whether it is an interglacial epoch like today or an ice age. What is crucial to the beginnings of an ice pack is whether the snow can survive through a summer, so it can begin to accumulate year after year. If, during the northern summer, the Earth is comparatively far from the Sun in its eccentric orbit, and if the axis is tilted less than usual, the summer may be a bit cooler and the snows may mount up. But if the Earth is near its closest distance to the Sun during a Canadian summer, and the tilt is greater, the Arctic Circle will extend farther south and the midnight Sun will last longer. Winter snows would be hard-pressed to survive such month-long heatings from the comparatively nearby Sun.

The astronomical changes seem very small—so small, in fact, that Milankovitch's contemporaries gave short shrift to his ideas.

More recently, however, precise measurement techniques have been developed that prove he was at least mostly right. Oxygen comes in two isotopes. The most common form, called "oxygen-16," has eight protons and eight neutrons in its nucleus. A rarer, heavier form, "oxygen-18," also has eight protons but ten neutrons. Water containing oxygen-16 preferentially evaporates from the oceans, so when it snows out and forms glaciers on the continents, the ocean is left enriched in the oxygen-18 that has stayed behind. Whether enriched or not, oceanic oxygen is incorporated in the shells of marine organisms that accumulate on the ocean floor. By charting the ups and downs of oxygen-isotope ratios, measured from seafloor sediments of different ages, scientists can thus infer the amount of ice stored in continental glaciers throughout geological history. Mathematical analysis of the data shows evidence that all three of the astronomical periods have affected the ice ages.

It is unsettling that such minor variations in sunlight—only a few tenths of a percent in the global average—could have such a profound effect on our planet's climate. But Milankovitch's insight about the actual climate-control process, involving 10% variations in sunlight at high northern latitudes, explains why the effects are so large. Still more alarming, however, is some newer chemical evidence that we still don't understand: One might expect that the slowly varying oscillations in the Earth's astronomical parameters would lead to slowly varying oscillations in climate, but that seems not to be true. Recent measurements of CO_2 trapped in Greenland ice sheets suggest that the CO_2 content of the Earth's atmosphere can change suddenly, and unpredictably, in only a few hundred years. It may be that the Earth's climate mimics an automobile driving down a rutted road, following one groove for a while, then abruptly crossing a threshold into another temporarily stable groove.

The end of the last ice age presents an example of a sudden oscillation between two quasi-stable states, one which affected prehistoric people. It had seemed that the ice age was over about 13,000 years ago, and that life was returning to normal, when the ice age suddenly resumed 11,000 years ago in Europe and lasted for another thousand years before abruptly ending again. There have been speculations about complex processes that may be responsible, involving details of the ice-sheet retreat and deep oceanic currents, but we do not understand them yet. We need a modern Milankovitch to provide us more insight.

The lack of scientific understanding about climate change is particularly disturbing today. Were we in the midst of an ice age, we could afford to burn our coal and oil for some time to come. The warming might well end the ice age, with its accompanying good and bad effects. (Ending an ice age might sound wholly good, but civilization would have adjusted to it, and *any* sudden change in the status quo can upset society's equilibrium.) But for a long time, such warming would be within the normal range of the glacial-to-interglacial cycle recorded in the geological record, so we could rest assured that no runaway greenhouse would happen. In contrast, today we sit at the peak of a warm, interglacial epoch. It has never been significantly warmer, and there has never been so much CO_2 in the atmosphere, since this cycle of ice ages began. Now any future warmings will push us into less well understood territory. To be sure, we may have a century or centuries to go before we exceed the heat of the Age of Dinosaurs; perhaps natural counterbalancing forces will keep us from ever getting that warm. But the geological record of tens of millions of years ago is more difficult to study than the recent cycle of ice ages. Just because it was great for the dinosaurs 100 million years ago, it might be awful for us (consider sea level a thousand feet higher than today!). And, for all we know, the Cretaceous may have been near the threshold for a runaway greenhouse, and changes set in motion today might not be reversible in time before our planet found itself rushing headlong toward a venusian conclusion. By contributing to the increase in CO_2, we are engaged in a risky experiment with the only planet we have.

These are the worries of pessimists—some would say realists—who emphasize the uncertainties in our knowledge of climate changes. On the other hand, many scientists who are trying to understand climate change are less worried. Based on their research during the past decade, they think they may understand some of the major elements of climate change. And they are inclined to believe that the Earth's climate is inherently stable against a runaway greenhouse. The stability won't protect us from the warming that has been forecast for the 21st century, but they believe it should protect our planet from becoming another Venus. For example, as the atmosphere warms, evaporation increases, so cloud cover increases. The brilliant clouds reflect the sunlight, helping to stabilize the temperature increases. Also, there is more rain, which more rapidly dissolves CO_2, weathers rocks to form bicarbonate ions, and washes the ions into the sea, where they are eventually buried in seafloor sediments, thus reducing CO_2 in the environment and working against the greenhouse. These are just two of several balancing factors that affect the climate. There are others. But until more research is done, we cannot be sure that instability won't rear its ugly head as civilization propels itself into a 21st-century climate not witnessed for the last million years or more.

* * *

Normal molecules of oxygen are composed of two atoms of oxygen. Ozone molecules have three oxygen atoms. The difference is profound. We require oxygen to breathe. But when we breathe ozone on a hot summer day, sitting in a traffic jam, our chest tightens in pain. There is another important, but beneficial, difference: A little bit of ozone is totally opaque to dangerous ultraviolet light. Ultraviolet rays, which are "more violet" than violet in the spectrum of the Sun, are especially energetic and destructive. They cause skin cancer, cataracts, and damage to our immune system. Human beings, like most life on Earth, evolved under atmospheric conditions in which ozone protected the Earth's surface from harmful ultraviolet, so we did not evolve any protective shells. That makes us dependent on the tiny amount of ozone that resides in the stratosphere, about 8 to 15 miles up. Remove the ozone, and we must stay inside, underground, or underwater . . . or else we die. Remove the ozone and many crops would suffer. That is why reports of the ever-widening Antarctic ozone hole are so threatening.

After nearly two decades of controversy, scientific proof has become overwhelming that the growing Antarctic ozone hole is due to chlorofluorocarbons (CFCs) such as DuPont's Freon. We are threatened with skin cancer thanks to refrigerants, aerosol spray-can propellants, plastic foams, and other miracle products of the chemical industry. The DuPont Company was recently the first in its industry to acknowledge the scientific conclusions; it announced plans to discontinue production of CFCs, but not until the turn of the century. In the interim, malignant melanoma will continue to kill.

Concern about our planet's ozone shield is not new. Nearly two decades ago, in the infancy of the environmental movement, technologists were proposing fleets of supersonic transport airplanes (the SST). Opponents feared they would threaten stratospheric ozone. The United States abandoned its plans, and in Europe, only a few Concordes and Soviet SSTs were built. A little later, atmospheric chemists discovered that chlorofluorocarbons manufactured for aerosol can propellants, and discarded when the cans were "empty," would leak their CFC contents into the air; the gases would ultimately find their way to the stratosphere and destroy ozone. The same chemical inertness that made the compounds so attractive and nontoxic for consumer uses are, ironically, responsible for their destructiveness in the upper atmosphere. Since they are so nonreactive, they survive for long times—decades to a century or more—in the lower atmosphere, which enables them ultimately to filter up to the ozone layer. Once above the ozone, they are attacked and destroyed by the dangerous ultraviolet rays, which break them down into their constituent atoms: chlorine, fluorine, and carbon. Chlorine attacks ozone extremely effectively. In essence, an atom of chlorine breaks apart ozone, causing it to recombine rapidly into ordinary oxygen; that, in turn, liberates the chlorine atom, which goes on to destroy another ozone molecule, and then another, and so on. Each chlorine atom derived from CFCs is estimated to be capable of destroying hundreds of thousands of ozone molecules each second. To make matters worse, there isn't very much ozone to destroy. Ozone, which itself is formed by ultraviolet destruction of ordinary oxygen in the stratosphere, constitutes only one millionth of the Earth's atmospheric gases.

In reaction to the public outcry of the mid-1970s, CFC propellants were banned in the United States and by other countries. Substitutes such as propane and isobutane, not to mention simple pump valves, were found to satisfy consumer demand. But when more critical products of the chemical industry were threatened, actions were postponed, committees were formed, and endless debates ensued. While CFCs continued to find their way into the stratosphere, scientists debated whether they were destroying a lot of ozone, or only a little. One report would say the threat was exaggerated, the next would refute the dismissal, and so on. The public, and officials in the laissez-faire Reagan Administration, lost interest and took a wait-and-see approach. President Reagan's controversial EPA administrator, Anne Gorsuch Burford, joked about the time, "a few years back, when the big news was fluorocarbons that supposedly threatened the ozone layer."

Everything began to come into focus in 1985. Theoretical arguments aside, reinterpretation of old satellite data and soundings from Halley Bay in Antarctica showed conclusively that ozone was substantially gone each spring in a continent-sized area centered on Antarctica. Subsequent measurements have shown that the "hole" has gotten emptier of ozone (down 50%), wider in extent, and is longer lasting. More recently, there have been reports of ozone deficits over the northern Arctic as well as over more populated parts of the world. Finally, the "smoking gun" surfaced: Chemicals intimately involved in the theoretical ozone-eating reactions were identified in the ozone hole, conclusive proof that CFCs are responsible. In early 1988, NASA's Ozone Trends Panel released its indictment: "The weight of evidence strongly indicates that man-made chlorine species [i.e. CFCs] are primarily responsible for the observed decrease in ozone within the polar vortex."

Unusual high-altitude polar clouds apparently help the chlorine-induced reactions to destroy ozone. But ozone depletion is not limited to polar regions. It is now thought that ozone has decreased in the last decade by perhaps 2.5% over the whole planet, exclusive of the polar regions. That slight loss—which is nevertheless larger than scientists had predicted—could lead to as much as a 10% increase in the rate of skin cancer. With ozone losses of 95% or more in the middle of the spreading ozone hole, it is all the more imperative that we stop experimenting with our fragile stratosphere. Invented just 60 years ago, CFCs are now produced and consumed at a rate of 3 billion pounds per year. Those already released into the atmosphere may be destroying ozone well into the 22nd century. With some luck, the effects could remain concentrated in the uninhabited arctic regions of our planet. But even the much lesser effects at lower latitudes are potentially disastrous. To make matters worse, CFCs may be second only to CO_2 in their ability to enhance the greenhouse effect.

The risks of tampering with the stratosphere were clear in the early 1970s. Even now, with the ozone hole staring us in the face, our political institutions seem to be incapable of dealing decisively and quickly with this proven hazard. The damage done from 60 years' use of CFCs—most of it produced since the first warnings were sounded—will last for generations. It makes one doubt if human beings are truly capable of politically controlling the dangerous side effects of the wonders that inventors create.

The tentative political response to the ozone hole discovery provides little comfort that the less sensational, more slowly developing greenhouse effect will be seriously addressed. After all, the CFC miracles of modern chemistry mostly serve the conveniences of the

Artist's view of the Antarctic ozone hole, and some of the consumer products that contribute to its growth. (Lynette R. Cook)

world's more affluent societies (air-conditioning, throwaway packaging, cleaners for electronic circuit boards). Furthermore, most CFCs are readily replaceable by nondestructive compounds. It will certainly be more difficult for the nations of the world to agree to stifle the consumption of fossil fuels, responsible for our warming trend. Society's basic need for cheap energy is much more fundamental, integrally related to our lives, and less easily replaced than are CFCs.

What must we do to reverse the destruction of the ozone layer, to halt the warming, to protect our world? We can begin by setting an example in how we lead our individual lives and by raising the consciousness of others to more harmonious ways of living with our environment. But the eventual solution must involve international political agreements; they must be strong enough to reign in the powerful economic forces that motivate industry and consumers alike to achieve short-term advantages at the expense of our collective, longer-term future. As citizens of one nation, we can attempt to persuade our own political leaders to act responsibly. But much of the environmental threat comes from the rapid industrialization of the heavily indebted Third World. For example, the most significant destruction of forests is being done in the tropics, in countries like burgeoning Brazil. It will take unprecedented international cooperation, including sacrifices by wealthier nations like our own, to reverse the alarming trends that scientists have been discovering.

One or more of the environmental dangers we have discussed could lead to a truly epochal catastrophe for our native world. It wouldn't happen in an instant, like an asteroidal collision. But on a cosmic time scale it could be swift, indeed, and just as devastating, if we are unable to respond constructively fast enough. As a culmination to thousands of years of history, human civilization has in the last few decades become a truly planetary force. In only a few more generations we will know whether the destructive potential of that force can be overcome by the innate intelligence and goodwill that are the most precious hallmarks of our species. The fate of the Earth, as the only abode of life in the solar system, may hang in the balance.

CHAPTER 17

Death of the Sun

Nothing lasts forever. Like all the rest of Nature, stars and their planetary systems experience a cycle of birth and death. Life has existed on Earth for nearly 4 billion years, and if we are lucky enough to escape from some of the more horrendous catastrophes we have described, life could survive for billions of years more. Ultimately, however, our time will pass. Like the generations of stars and planets that preceded it, our Sun will exhaust its nuclear fuel and die. The manner of its dying is the subject of this chapter: Will the Sun and Earth go out with a bang, or a whimper?

The Sun must die because it will run out of fuel and no longer be able to generate the energy that provides light and heat to its planetary system. The idea that the Sun has a finite lifetime is a couple of centuries old, but the problem has been to estimate just what this lifetime is. To do so, we must understand the source of the Sun's energy.

The science of astrophysics was born in the middle of the last century, when physical concepts of the nature of matter and energy were first applied to celestial systems. Once astronomers had determined that the Sun was a giant globe of incandescent gas, radiating energy into space at the rate of 100 billion megawatts, they began to inquire after the source of this energy. As long as the Earth and Sun had been thought to have an age of only a few thousand years, the question never arose, since a hot mass of gas the size of the Sun contains enough heat energy to shine undiminished for hundreds of thousands of years, even without a new source to replenish the energy radiated into space. But by the early 19th century uniformitarian geologists were discussing an Earth that must be at least several millions of years old. About a century ago, pioneering British physicist James Thompson, also known as Lord Kelvin, calculated the Earth's age to be about 30 million years, based on the time required for the planet to cool to its present temperature from an initially molten state. We now know this cooling-time calculation leads to a meaningless result, since the Earth's internal temperature is maintained by energy released from natural radioactivity, but radioactivity was not discovered until 1899. Thus both the vast time scales being suggested by geologists and this calculation of the cooling time for the Earth challenged 19th-century astrophysicists to find an explanation for the energy source of the Sun.

A natural possibility was that the Sun was literally burning, deriving its energy from the same kind of chemical reactions that power the furnaces and forges of industry on Earth. The mass of the Sun is about that of 200,000 Earths, and it is a simple matter to calculate the chemical energy that would be released if it consisted, for example, of coal and oxygen that could combine through ordinary combustion. Actually, the maximum chemical energy would be released if the Sun consisted entirely of hydrogen and oxygen, the same chemicals used to power the Space Shuttle and other advanced rockets. How long could the Sun last if its energy source were the combustion of hydrogen and oxygen? The answer is only about 3000 years, no longer than the simple cooling time for a hot sphere of gas. Besides, it was becoming clear that, while the Sun contains a great deal of hydrogen, it is very deficient in oxygen, so the idea of a giant chemical furnace based on oxidation reactions was not very attractive.

Physics came to the rescue for a time with the suggestion that the Sun might derive its energy simply from shrinking. Any falling object gains kinetic energy, a fact we relied on earlier when we discussed the energetics of impact cratering. In the Sun's gigantic gravitational field, impacting bodies gain tremendous energy, which they impart to the Sun when they strike. The same physical principle tells us that any shrinkage of the Sun will generate heat by an analogous mechanism, since shrinkage involves the "falling" of the outer parts toward the interior. Knowing the Sun's mass, physicists like Lord Kelvin calculated that it could shine this way for about 50 million years, if it had started as a much larger body and had been steadily shrinking in size at a rate of a few feet per year. This seemed to provide an entirely satisfactory explanation for the Sun's energy, and it suggested that the Sun could continue to shine by the same mechanism for several tens of millions of years into the future. The solution was short-lived, however, for geologists early in the 20th century used the newly discovered concepts of radioactivity to estimate the Earth's age as hundreds of millions, and then billions, of years. And in 1925 astronomer Edwin Hubble startled the world with his discovery of the expansion of the universe, which he was able to calculate began at least 2 billion years ago (a value later repeatedly revised up to the present figure of about 15 billion years). Thus once again astrophysicists were challenged to come up with a satisfactory explanation for the energy sources of the Sun and stars.

Albert Einstein had provided the essential clue in 1905, with his famous conclusion that matter and energy are equivalent. This result from the special theory of relativity, familiar to us from the equation $E = mc^2$, indicated that matter itself contains an amount of energy millions of times greater than its chemical energy or the energy that could be released by shrinking of the Sun or stars. If relativity were correct, the universe contained an almost limitless source of energy. But how could this energy be released and utilized?

During the 1930s a number of astrophysicists worked on the problem of identifying the nuclear reactions that could power the Sun and stars by the conversion of matter to energy. They calculated that the most efficient sources of nuclear energy should result from the fusion of four atoms of hydrogen into one atom of helium. When such a fusion reaction takes place, the helium atom ends up short of mass, compared with the four hydrogen atoms that combined to form it. The missing mass, amounting to 0.7%, is converted to energy according to Einstein's famous equation. Over a period of a few years, the essential reactions were worked out. One of those involved in this work was the physicist George Gamow, who had emigrated to the United States from Russia. Colleagues tell the story that one day he had an insight that permitted him to solve a critical problem that had been holding up the calculations, and he realized at once that he had identified a series of reactions that really would work in the interiors of stars. He is supposed to have taken his girlfriend out that night and parked under the stars, where he confided that he was the only person in the world who knew what made the stars shine!

The point of this history lesson is to teach us how to calculate the lifetime of the Sun. Like other stars, it derives its energy from the fusion of hydrogen to helium in its deep interior, at a temperature of millions of degrees. To maintain its present luminosity, 600 million tons of hydrogen are converted to helium each second. Luckily for us, most of the material in the Sun (75% by mass) is hydrogen, so there is plenty of fuel to stock this nuclear furnace. A simple calculation shows that the time to exhaust the hydrogen initially contained in the Sun is a bit more than 10 billion years.

The Sun and the solar system are known to be 4.5 billion years old, so our star is about halfway through its lifetime. There is enough hydrogen in the Sun to keep it shining at the present rate for 5 to 8 billion years more. We can therefore all heave a great sigh of relief. But hold on! How do we know that the Sun will remain stable for all of this time? What happens as the hydrogen is depleted and the solar structure readjusts to compensate for the changing composition of the core? Could the Sun explode? To answer such questions we must go far beyond the simple identification of the Sun's energy source; we must delve into the arcane calculations of stellar evolution.

* * *

Among the triumphs of modern science is its ability, aided by the development of high-speed computers, to follow the life stories of the stars. Stellar evolution takes place on billion-year time scales, so we rarely have the opportunity to see an individual star pass from one stage of its life to another. Yet we can model the processes that take place in a star's interior and calculate the effects on its structure and luminosity as it slowly uses up its nuclear fuel. The results of these evolutionary calculations can in turn be checked against the observations that astronomers make of thousands of different stars, each caught in a different moment of its life cycle. If we find only a few stars in stage A, but hundreds in stage B, we may assume that typical stars pass very rapidly through stage A, while stage B represents a fairly stable configuration for an evolving star.

Stars are born out of condensations in the clouds of gas and dust that are scattered throughout the disk of our galaxy. These clouds contain not just primordial hydrogen and helium, but also all of the heavier elements, which were themselves synthesized in the interiors of earlier generations of stars. The Sun, the solar system, and ourselves, are composed of starstuff, the product of 10 billion years of galactic evolution. By the time 4.5 billion years ago when our Sun formed from one of these clouds of dust and gas, there had accumulated sufficient heavier elements to permit the formation of the planets, comets, and other members of the planetary system.

Initially, the Sun heated up as it derived energy from shrinkage of the cloud in which it formed, just as had been calculated in the last century by Lord Kelvin. Eventually it became so hot and dense that the fusion of hydrogen to helium could begin in its core. This moment when the nuclear fires were kindled marks the true birth of the Sun as a star, shining from its own internally generated energy. As this nuclear energy began to push outward from the Sun's core, it counteracted the collapse, and the Sun stopped shrinking. Within a few million years, the Sun stabilized in a configuration in which the internally generated energy exactly balanced the heat and light radiated from the surface. Astronomers describe any stable star powered by hydrogen fusion as being on the "main sequence." Because this is a fairly stable configuration during a star's evolution, about 90% of the stars in our sky fit into this category. As we know, the Sun began its main-sequence lifetime about 4.5 billion years ago.

Any star can be described as existing in a state of dynamic tension between two opposing forces. On the one hand, gravity is constantly trying to pull the star more tightly together. Resisting this tendency toward collapse are the high interior temperature and pressure, maintained by internal nuclear reactions. The Sun, like most stars, has achieved an exact balance between these forces of contraction and expansion, which is lucky for us, since life would be difficult or impossible on a planet orbiting an unstable or pulsating star.

As the hydrogen in its core is gradually used up, however, a star slowly adjusts itself to its changing internal composition. Calculations show that it expands slightly and increases the rate at which hydrogen fusion reactions take place in its interior; over several billion years, the luminosity of the Sun goes up about 50%. Other, more subtle, changes may also take place in the circulation of gases in the solar interior. But on the whole the Sun remains roughly the same for at least 8 billion years, or until more than half of its hydrogen has been converted to helium. At that point it will have assumed a layered structure, with a large core of nearly pure helium, surrounded by a shell in which hydrogen fusion is still taking place, surrounded in turn by an atmosphere that retains the original composition with which the Sun was formed.

When the helium core gets large enough, the Sun loses the rock-like stability that has characterized its long main-sequence lifetime. The age-old battle between the forces of expansion and contraction enters a new phase, as the enlarged core finds itself unable to support its own weight. The Sun is about to increase dramatically in both

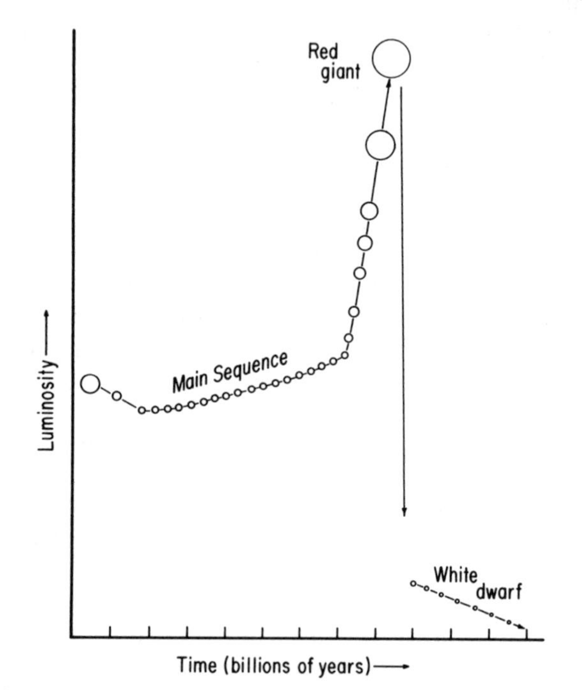

Evolution of a star like our Sun. It evolves in size and luminosity from the main sequence to become a red giant, and ultimately a white dwarf.

luminosity and size, becoming what astronomers call a "red giant star." In the interior, gravity becomes dominant, and the helium core contracts to generate the energy required to maintain its high temperature and support the weight of the overlying layers of gas. The addition of this new source of gravitational energy from the shrinking core increases the luminosity of the Sun substantially, and the added energy streaming from the core pushes out the atmosphere to remarkable dimensions. In the case of the Sun, the swelling atmosphere will engulf the planets one by one: first Mercury, then Venus, then Earth, and finally Mars. But long before it disappears into the fiery gases of the distended Sun, our planet will have ceased to be habitable.

The rising luminosity of the Sun will raise the temperatures on all of the planets. The ice caps and subterranean ices of Mars will melt and rivers may flow again on its desiccated surface, perhaps making it (again?) a habitable world. Higher temperatures will lead to expansion and ultimately loss of the atmosphere of Venus, while the surface of Mercury will be so hot that the rock will become plastic and the ancient craters will disappear forever. On our own planet, temperatures will increase year by year, and century by century, as the distended Sun pours out more and more energy. Life, if it has survived to this period billions of years in the future, will be extinguished by the terrible heat. The oceans will boil, creating a huge atmosphere of water vapor, but quickly this atmosphere will itself escape into surrounding space. By the time the outer envelope of the Sun begins to brush against the Earth, our planet will have been reduced to a cosmic cinder, already softening as its rocks ooze and melt and its continents slide slowly into the empty basins that once held the oceans.

A red-giant star is so large that its atmosphere is reduced to very low density, lower than that of the upper atmosphere of the Earth today. The density is so low that Earth's motion will hardly be affected as our planet, already burned to a crisp, begins to be consumed by the swelling Sun. It will continue in its orbit, much as an artificial satellite circles the Earth within the tenuous uppermost reaches of our atmosphere. Just as the orbit of an artificial satellite eventually decays from air friction, however, so will the Earth-cinder gradually spiral into the atmosphere of the red-giant Sun. The process will probably take tens of thousands of years, but eventually our planet will be vaporized and become a part of the Sun itself. This is the ultimate fate of the atoms that constitute our own bodies: to rejoin the fiery gases of the Sun.

The accuracy of this prediction of the end of the Earth is thought to be high. There is little question that the Sun will exhaust its hydrogen or that it will swell into a red giant large enough to engulf the Earth and probably Mars. It is equally certain that at present the Sun has not reached the end of its main-sequence lifetime, nor is it likely to do so for billions of years. Also, the science-fiction scenario of the Sun "blowing up" as a supernova is not among the cosmic catastrophes that should concern us. Still, it would be foolish to assume that our calculations are infallible, or that the Sun may not yet have surprises for us.

* * *

One of the mysteries of the Sun is associated with the increase in luminosity that is predicted during its main-sequence lifetime. Model calculations indicate that the Sun should be putting out about 30% more energy today than it did 4 billion years ago, when life originated on Earth. The problem arises because we know that a 30% decrease in solar luminosity, if it occurred today, would lead to the rapid freezing of the oceans and plunge our planet into a terminal ice age. Yet the history of life on Earth tells us that the oceans of liquid water have been present for at least 4 billion years, and that the general climate over at least the past billion years has been quite similar to that of today. It appears, therefore, that the Earth must have compensated in some way for changes in solar luminosity so as to maintain about the same mild climate. How is this possible?

One way that the Earth could have sustained a temperature above the freezing point of water in spite of the lower former luminosity of the Sun is by the greenhouse effect, which we discussed earlier. A substantially larger proportion of carbon dioxide in our planet's atmosphere could have maintained a surface temperature higher by tens of degrees, sufficient to maintain a mild climate in spite of a lower input of solar energy. Since we think that the Earth's original atmosphere was thicker than it is today as well as being greatly enriched in carbon dioxide, this explanation seems to make sense. But what a remarkable coincidence, that the declining carbon dioxide content of the atmosphere should have compensated so closely for the increasing luminosity of the Sun over a period of 4 billion years. This relationship is even more remarkable when we consider that it is life itself—specifically the development of green-plant photosynthesis—that has led to the decline in carbon dioxide and the concomitant increase in oxygen. It is as if the evolution of life conspired, through the changing composition of the atmosphere, to maintain the climate in which it could flourish. Many scientists believe that we are only beginning to explore the complex relationships that may exist between life—the biosphere—and the Earth's atmosphere and oceans.

Can this benign coupling between the animate and inanimate worlds continue to compensate for the slow brightening of the Sun? Already the Earth's greenhouse effect has been reduced by natural forces to a very low level, and it seems doubtful that much more could be done to resist warming. Instead, the rapid alteration of the environment precipitated by human activities such as the burning of fossil fuels and the destruction of the Earth's forests are now increasing the levels of carbon dioxide. By our own actions we seem determined to enhance the greenhouse effect, with catastrophic consequences, long before such a fate is forced on our planet by the slow increase in the Sun's luminosity.

The Sun might affect the climate of the Earth in other ways, all on time scales much shorter than those associated with evolutionary brightening. Suppose that, superimposed on this slow increase in luminosity, there are short-term variations in the energy reaching us from the Sun. We know that the Earth's climate can change from one century to the next. For example, there was a period of warmer temperatures about a thousand years ago associated with the great voyages of discovery by the Vikings, who settled in areas of Greenland and Newfoundland that are too frigid to support an agricultural economy today. On the other hand, there was a cooling in the 1600s, sometimes called the "Little Ice Age," reflected in European historical records (and some famous paintings) of skaters on the canals of Holland or London fairs held on the frozen Thames River, places that very rarely experience ice in today's world climate. These changes, which are clearly recorded in history, are probably the result of short-term variations in the brightness of the Sun.

During the past decade solar astronomers have become very interested in exploring the possible causes of these recent climatic variations. There is an apparent correlation between the Little Ice Age and a span from 1650 to 1710 during which no sunspots were seen, as deduced from astronomical records in both Europe and China. Sunspots themselves are tracers of a much more extensive set of solar phenomena often called simply "solar activity." In a well-documented 22-year cycle, the Sun's magnetic field flips from north to

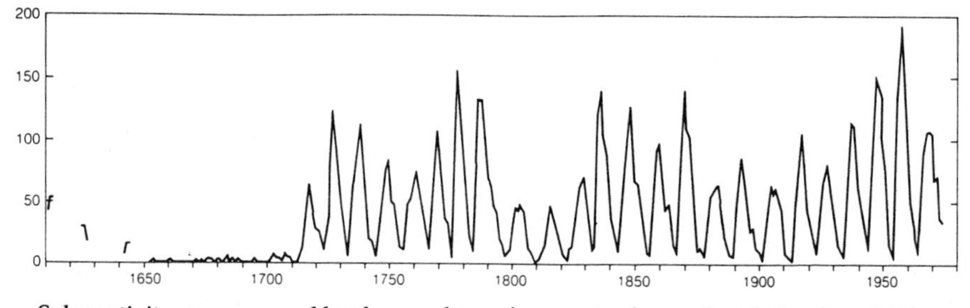

Solar activity, as measured by the numbers of sunspots observed each day from 1600 to the present. The absence of sunspots between 1645 and 1715 may be related to a worldwide drop in temperature at about the same time. (From *Realm of the Universe*, 4th ed., by G. Abell, D. Morrison, and S. Wolff, Saunders College Publishing, 1988)

south and back again. Accompanying this reversal of magnetic polarity are a host of other changes, including the extent and temperature of the Sun's outer atmosphere (the corona) and the solar wind of charged atomic particles that flows outward through the solar system. Deeper in the Sun's atmosphere, this magnetic cycle also influences the numbers and locations of sunspots (dark circular regions representing reduced atmospheric temperatures) and of the energetic explosions called "solar flares." All of these manifestations of solar activity have the potential to affect the Earth, through small variations in the solar luminosity as well as more directly through interactions of our planet with the outflowing solar wind. It is surely more than a coincidence that a century of greatly reduced solar activity, quite unlike anything seen since, was also the time of the most recent recorded dip in the temperature of the Earth.

Today orbiting satellite observatories are capable of measuring changes as small as 0.1% in the energy output of the Sun, and scientists are monitoring these variations closely. They know that a 1% variation in luminosity is sufficient to have major economic consequences on Earth. While 1% may not sound like much to us, that is enough to transform the fertile "breadbasket" grain-growing areas of the American high plains or the Soviet Ukraine into desert dust bowls, or to initiate large-scale melting of the Arctic ice pack. These are examples of how minor oscillations in the cosmos, far short of the catastrophes we have been discussing, may nevertheless tip a delicate ecological balance in a locale on Earth and trigger an economic catastrophe.

Other scientists are examining ice cores from Greenland and the Antarctic to obtain information on variations in Earth's climate extending over the past ten thousand years. But all of these studies are in their infancy. The fact is that we understand the solar magnetic and activity cycles very poorly, and their connections with terrestrial climate are even more tenuous. Small variations in the Sun, below the threshold of interest to most astronomers or astrophysicists, are events of very great interest to you and us. The best we can say is that life has survived on Earth for billions of years, and there are no indications of any major changes in climate over that period. But the current human population of the Earth is surviving at the margin in terms of agricultural and energy production, and it is sobering to realize how small a change in the Sun might precipitate a major economical and social upheaval on Earth.

CHAPTER 18

Supernovas: Nature's Most Violent Catastrophes

Until 1987, one fact about the stars had remained the same in every astronomy textbook ever published: In the past thousand years, only five supernovas had been visible to the naked eye, the last one dating from before the landing near Plymouth Rock. Now there are six. On February 23, 1987, a new star burst forth and gleamed in the southern skies. Although the star has now faded from view, it is still being chronicled by astronomers who never dreamed that they would be alive to witness so clearly the violent death of a massive star. Supernova SN1987A has given scientists new confidence that they really do understand much about these awesome events.

Old age is a troublesome time for a star. During most of its lifetime, a star achieves balance between the forces of gravity and the internal nuclear furnace that maintains its energy. But when the hydrogen that powers this nuclear furnace becomes exhausted, gravity dominates and the star becomes unstable. First it expands into a red giant, accompanied by gravitational contraction of its helium core, which can no longer generate energy from hydrogen fusion. For stars of moderate mass, like the Sun, the red-giant stage may be followed by a period in which the dying star sheds mass in the form of successive shells, explosively ejected. In spite of these paroxysms, however, such a star must ultimately give up the struggle against gravity. It can support itself only so long as it generates energy, either from thermonuclear fusion or by slow gravitational contraction. In the end, when all its energy sources have been exhausted, only a faint hulk is left: a white dwarf.

A white dwarf has the mass of the Sun compressed into the

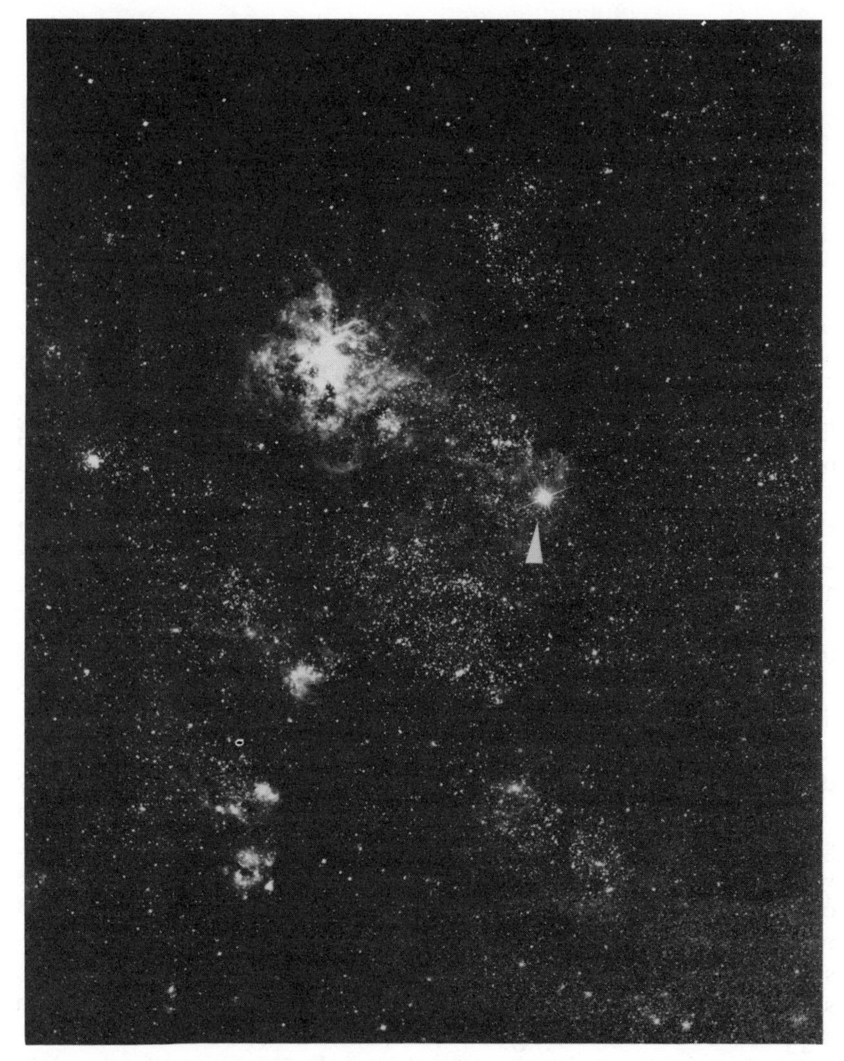

The supernova known to astronomers as SN1987A, which appeared in a nearby galaxy in 1987; the first naked-eye supernova in 300 years. (National Optical Astronomy Observatories)

much smaller volume of the Earth. It is extremely dense; a teaspoon would weigh a ton. In this highly compressed state, the white dwarf finally achieves equilibrium, with its weight supported against gravity by the strength of the atomic structure of the material itself, much as a building on Earth supports itself against collapse by the strength of the wood, brick, cement, or steel of which it is composed. Most stars are fated to end their lives as white dwarfs, which cool further and eventually cease to shine. Someday the Sun will be such a burned-out cinder, perhaps still orbited by the deteriorated remnants of its outer planetary system.

The key to a star's final fate lies in its mass. "The bigger they are, the harder they fall" is an old saying that applies very well to stellar evolution. The bigger or more massive a star, the more energy can be tapped by its gravitational collapse. But the bigger it is, also, the more difficult it is for the star to halt its collapse once it begins. The incredible weight that must be supported is simply too much for even the highly compressed matter in a white dwarf. Stars much more massive than our Sun cannot die quiet deaths; in fact, their agonies can generate the awesome cataclysms known as supernovas.

A supernova represents the ultimate cosmic catastrophe. In its terminal collapse, a massive star can shine for a few weeks with the light of 100 billion Suns. Yet even this brilliant flash represents but a small fraction of the energy that is released in such an explosion, as we now know from observations made of SN1987A.

* * *

The 1987 supernova helped to relieve a sense of frustration that astronomers had been accumulating for decades. Theoretical calculations, confirmed by observations of supernovas in other distant galaxies, indicated that once every 50 years or so, one of the 200 billion stars in the Milky Way galaxy should go supernova. Yet not one of these stellar spectaculars has been observed in our galaxy during the past 300 years.

The three earliest records of supernovas are to be found in the court archives of imperial China, a civilization that encouraged observations of the heavens during the intellectual dark ages that engulfed Europe for a thousand years before the Italian Renaissance. The Chinese recorded the appearance of these "guest stars" in the years 1006, 1054, and 1181 A.D. The first was reported to be nearly as bright as the full moon, yet there are no European records to confirm this remarkable apparition. The brightest supernova, first seen at the beginning of July 1054, remained visible in broad daylight for several weeks, and the Chinese continued to record its slowly declining brightness until it finally disappeared nearly two years later, in April 1056. It is very difficult to imagine that such obvious and long-lived celestial events should not have been recorded in Europe, but such is the case. Perhaps they were overlooked as a result of an Aristotelian superuniformitarian mind-set that refused to accept the possibility of change in the "perfect" celestial realm.

The two remaining supernovas in our galaxy were seen just at the birth of modern astronomy. The first, in the year 1572, was studied in detail by the greatest of the pretelescopic observers in Europe, the eccentric Danish nobleman Tycho Brahe, and it is sometimes called "Tycho's star." The second of these galactic supernovas, appearing in 1604, was of great interest to Galileo's German contemporary Johannes Kepler. Six years later Galileo invented the telescope, but in all of the time since, not one supernova has graced our night skies, although 20th-century astronomers photograph them regularly in other distant galaxies.

Given this absence of data, it is impressive that astronomers had developed any understanding of these violent stellar explosions. Only the optical outburst itself, in which the dying star explosively ejects much of its mass into surrounding space, can be studied in the faint extragalactic supernovas millions of light-years away. The precursor star is far too faint to have been photographed before the explosion, and in a few months the fading supernova sinks back below the measurement threshold of even the largest telescope. As a result, the explosion itself can be characterized, but there can be no direct evidence concerning the star either before or after its disruption, and little means to study the process that might trigger such a violent event.

Information about the aftermath of these explosions was derived from study of the remnants of galactic supernovas. Thanks to the anonymous Chinese astronomers, Tycho Brahe, and Johannes Kepler, we know where in the sky five such explosions occurred, and astronomers could therefore search in those locations for any unusual phenomena that might still provide clues to the events that took place centuries before. First they located the still-expanding shells of ejected gas from the explosions. Associated with these still-glowing gaseous remnants were surprisingly powerful sources of cosmic radio waves, so strong that it was certain there must be a large continuing injection of energy. What could the source of this energy be? Was there a faint remnant star as well as the expanding gas cloud, and if so, could this surviving fragment be the sought-for energy source?

* * *

The key to unlocking the mystery came in 1967, at a large radio telescope constructed by Antony Hewish of Cambridge University, England. The apparatus resembled an array of low wire clotheslines stretched out a few feet above the ground, connected to sensitive electronic detectors. In the summer, the astronomers brought in sheep to graze under the wires, since they were too low for a lawn mower. One of the projects at this radio observatory was a search for rapidly varying sources of cosmic radio energy. The effort was entrusted to a graduate research assistant, Jocelyn Bell, who made the crucial discovery for understanding the nature of supernovas.

What Jocelyn Bell discovered in the constellation of Vulpecula was a source of rapid, sharp, intense, and extremely regular pulses of radio energy, spaced at exactly 1.337 seconds. This phenomenon was so remarkable that for a time Bell and the other astronomers speculated that they might have picked up some sort of interstellar navigation beacon constructed by intelligent beings. Soon additional sources were discovered, similar to the first but each with its own regular period. One of these was located in the Crab Nebula, a still-expanding glowing celestial cloud (visible in backyard telescopes) at the location of the supernova of 1054; the period in this case was only about 0.033 second. Since then, more than a hundred of these pulsating radio sources have been discovered, most of them associated with supernova remnants like the Crab Nebula.

These objects were called "pulsars" because astronomers first thought they were stars that vibrated or pulsed. Theorists quickly realized, however, that it is not physically possible for any known type of star to vibrate with such short periods, often less than a second. The alternative idea was that the sources were spinning at these rates, but that too is impossible for known stars. Whatever these pulsars were, they must be smaller and denser than white dwarfs. Soon it was realized that astronomers were dealing with an entirely new kind of object, a star no larger than 10 miles in diameter, spinning at a rate of many times per second and generating an enormous amount of energy, which is transferred into the surrounding gas cloud.

The Crab Nebula is an expanding gas cloud that remains from the bright supernova of 1054. Inside the nebula is a rapidly spinning neutron star. (Palomar Observatory photo)

The only way a star can be compressed to such a small size is if its constituent atoms are crushed by pressure until they merge into a single object with the density of a neutron, one of the building blocks of the atomic nucleus. Such an object is called a "neutron star," and it has a billion times the density of a white dwarf. Thus pulsars are rapidly spinning neutron stars, and given the association of pulsars with supernova remnants, it must be that the final product of a supernova explosion is a neutron star.

To understand how a neutron star might be formed, let us return to the struggle between gravity and internal pressure that characterizes the life history of stars. For less massive stars like the Sun, the issue is resolved when they collapse into stable white dwarfs. A problem arises, however, for stars that are much more massive than the Sun. Since the 1930s, astronomers have realized that the maximum mass for a white dwarf is only about 1.4 times the mass of the Sun; larger than this, a white dwarf simply cannot support its own weight. It must collapse further, crushing the atoms of which it is made until the collapse is ultimately resisted by the super strength of atomic nuclei. Thus the existence of neutron stars had been predicted from theory decades before they were discovered by Jocelyn Bell.

The collapse to neutron densities is inevitable if the core of an old, evolved star is more massive than 1.4 solar masses. This collapse, taking place after all nuclear sources of energy are exhausted, is much more violent than the collapse to a white dwarf, and it releases prodigious quantities of gravitational energy. A part of this gravitational energy, in turn, blows away the atmosphere of the star and gives rise to the observed supernova explosion; theory also predicted that even more energy would be emitted in the form of the massless subatomic particles called "neutrinos." Many other nuclear reactions also take place during the brief seconds of the core collapse. This in outline is the picture of the supernova process that emerged from theoretical considerations, even though no star had ever been observed in a presupernova state, no supernova neutrinos had ever been detected, and even the explosion process itself was not well observed due to the faintness of supernovas in galaxies far beyond our own Milky Way. Against this backdrop, SN1987A appeared, not quite in our own galaxy, but in the next closest one: in the nearby companion galaxy known as the Large Magellanic Cloud.

* * *

The Clouds of Magellan are two smallish companion galaxies to our own Milky Way spiral. Visible from the Earth's southern hemisphere, they appear in the sky like detached wisps of the Milky Way, each with a diameter several times that of the full moon. Unfortunately, they are so near the south celestial pole that neither can be seen from Europe or North America. They are called Magellanic Clouds because their existence was first reported in Europe by members of Magellan's 16th-century expedition to circumnavigate the globe. This southern location has proved to be an inconvenience for astronomers, since most observatories are located in the northern hemisphere. In fact, desire to study the Magellanic Clouds was a major motivation for the development of southern observatories, principally the Cerro Tololo Interamerican Observatory in Chile supported by the United States National Science Foundation, the nearby European Southern Observatory, also in the Chilean Andes, and the Anglo-Australian Observatory in Australia.

It was astronomers at Cerro Tololo and another southern observatory, Las Campanãs, who first spotted the "new star" in the Large Magellanic Cloud on the night of February 24, 1987. This rapidly brightening object was already visible to the naked eye at its discovery. Within a matter of hours, the word of this event—the first supernova visible without a telescope since 1604—flashed around the world, and normal work at the southern observatories ceased as all telescopes were focused on this unique celestial spectacular. That same day, controllers at NASA's Goddard Space Flight Center near Washington redirected the only operating American space observatory, the International Ultraviolet Explorer, toward the supernova. Astronomers in the Soviet Union eagerly awaited the checkout of their 20-ton space astrophysics observatory Kvant, which had just been launched to link up with the Soviet orbiting space station Mir, where it would provide a long-term capability to observe the supernova's changes from above the atmosphere.

As astronomers began on February 24 to investigate the brilliant new light in the southern sky, however, the most important data on the supernova had already been recorded, by "telescopes" very different from those we usually think of on the Earth or in orbit. Conventional astronomical equipment measured the expanding envelope of gases blown off the star. But much more fundamental is information on the underlying collapse of the star's core to form a neutron star, and this information had reached the Earth the previous day in the form of neutrinos.

Neutrinos are the most elusive of subatomic particles, because they hardly interact at all with matter. To detect them, physicists have constructed huge underground tanks of water, isolated from any disturbance by ordinary cosmic rays, which are trapped within the first few feet of the Earth's surface. Here instruments of extraordinary sensitivity wait hour after hour, and day after day, to record the tiny fleeting flash of light that can be produced when an occasional neutrino is absorbed by the water. One of these "neutrino telescopes," constructed by the Irvine–Michigan–Brookhaven consortium, is located in a deep mine underneath Lake Erie; another is buried below the town of Kamioka in Japan. Both were patiently waiting on February 23, 1987, when at exactly 7 hours, 35 minutes, 42 seconds Greenwich Mean Time (GMT) the pulse of neutrinos from the collapsing core of the supernova passed through the Earth. During the fleeting 5 seconds that this pulse lasted, the Lake Erie instruments detected 6 of these neutrinos, and the Kamioka instrument detected 12. Of course, the supernova was not itself visible from either the United States or Japan, but no matter; the neutrinos shot right through the Earth and entered the detectors from below. This remarkable property of neutrinos, to penetrate solid objects as large as the Earth, is what makes them so hard to capture, but also endows them with their unique value for supernova studies, since they are able to pass unhindered from the collapsing stellar core through all of the overlying layers of the star directly to us. These neutrinos thus provided a window into the center of a dying star.

The detection of the neutrino pulse was unique to SN1987A. If this supernova had taken place a few years earlier, there would have been no neutrino telescopes on Earth to record the event. The other unique aspect, compared with all other supernovas, was astronomers' ability to identify the nature of the star before it exploded. The parent object of the Large Magellenic Cloud supernova turned out to be a massive and highly luminous blue-white star called Sanduleak −69 202, which had been well observed over the years. Its distance is 175,000 light-years. Although the star had seemed quite stable, its mass was known to be 15 times the mass of the Sun, far too large to permit it a quiet old age as a white dwarf. Stars this massive are destined from birth to follow a one-way road to violent destruction. Combining the observed facts with an array of powerful theoretical models, astronomers can now trace with some confidence the death of Sanduleak −69 202.

* * *

Long before there were astronomers and telescopes on Earth, Sanduleak −69 202 had exhausted the hydrogen in its core and evolved into a red giant. For a star of such a large mass, the contraction and heating of the core leads to nuclear reactions more complex than the simple fusion of hydrogen to helium. First the helium in the core reacts to produce carbon and oxygen, generating enough heat to stabilize the star for a time. It shrinks and retreats from red-giantism, looking for a time like an ordinary luminous main-sequence star. Then this new energy source is used up, and core contraction begins again, pausing only when further nuclear reactions that feed on carbon and oxygen are initiated. The star may expand and contract several times as it rapidly works its way through the periodic table of elements, synthesizing new atoms as its luminosity and central temperature continue to rise. The star assumes a layered structure, each region having a different elemental composition, like the layers of an onion. This process can continue only so far, however. When the inner core is converted to iron, it will have used up its last gasp of nuclear energy, since any further synthesis of heavier elements would absorb energy rather than release it.

At this stage, the iron core, with a mass about twice the mass of the Sun and a diameter similar to that of the Earth, begins its final contraction. At first it shrinks slowly, generating gravitational energy to feed the star's unquenchable appetite. Then suddenly, at a temperature of 50 billion degrees Fahrenheit, a reaction starts that releases 99% of its energy in the form of neutrinos. Unlike the gamma rays and other forms of energy produced up to this point, the neutrinos escape from the core directly, moving at the speed of light. The result is equivalent to withdrawing all support from the overlying layers of the core, which proceeds to collapse almost instantly, generating a thousand times as much energy in the ensuing 5 seconds as the Sun will in its entire 10-billion-year lifetime. The iron core becomes a neutron star, and the rest of the star is about to be torn apart.

The small fraction of energy released in forms other than neutrinos generates a rapidly rising shock wave that spreads up to the surface of the star, which it reaches in a little over an hour. On the way, the shock wave violently heats the layers of the star, generating new nuclear reactions; one of these converts silicon to a short-lived radioactive form of nickel. All of these newly synthesized elements are ejected into space along with the remains of the star's upper layers. The explosion cloud, with an initial temperature of 100,000° F and an expansion speed of 15,000 miles/sec, can be observed within a matter of hours by distant astronomers with ordinary optical telescopes.

Although the core collapse of SN1987A actually happened 175,000 years ago, as seen from the Earth it took place at exactly 7 hr, 35 min, 42 sec GMT on February 23, when the neutrinos, arriving at the speed of light, burst through the Earth. About 4 hours later the first photo of the rapidly brightening star was taken by automatic cameras in Australia. A day after the initial collapse, the expanding shell of hot gas rendered the star visible to the naked eye, and it was discovered. It continued to brighten slowly for another 80 days, powered in part by the decay of radioactive nickel to a radioactive form of cobalt, reaching its maximum brightness in mid-May. Then, as the shell of ejected gas cooled and became transparent, the supernova began its long decline. After another month, however, the decline smoothed off to a constant rate, characterized by a drop just equaling the half-life for the radioactive decay of cobalt to iron. Most of the visible energy of the supernova was by then derived from the radioactive elements so violently synthesized in the explosion.

In August, 249 days after the explosion, this conclusion was confirmed directly by observations of gamma rays produced by this radioactive reaction. Shortly after the initial detection from a Japanese satellite, additional gamma rays and X rays from the supernova were picked up by the Soviet Kvant astrophysics module on the Mir space station. One NASA scientist called these observations "the confirmation of 40 years of computing models of stellar interiors. It is the first time we have seen direct evidence of newly processed material inside a star."

* * *

SN1987A, 175,000 light-years distant in the Large Magellanic Cloud, is of extraordinary interest to astronomers, but of no consequence to daily life on Earth. The same could be said for most supernovas in our own Milky Way galaxy, one of which could appear at any time. These stellar extravaganzas would assume more than academic interest, however, if one of them took place close to the Earth. Suppose, for example, that Sirius, the brightest star in the sky at a distance of just 8 light-years, should become a supernova. At its peak, it would shine almost as brightly as the Sun, turning night into day. The Earth would be flooded with X rays and gamma rays, followed a few months later by cosmic rays produced in the explosion. This energetic radiation would alter our atmosphere, destroying the ozone layer that protects us from solar ultraviolet light; at the same time the radiation from the supernova itself would induce serious genetic damage in living tissues. In all probability, the Earth would be rendered uninhabitable. Even if the supernova were ten times farther away than Sirius, the damage to living things would be considerable. Some people suggested, before the recent association of mass extinctions with impacts, that these great dyings might have been triggered by relatively nearby supernova explosions.

It is fairly simple to look around us and see if there are any stars in the solar neighborhood that are due to become supernovas. Sirius is the most massive star within 20 light-years of the Earth, and we are pleased to report that its mass is only twice that of the Sun: not enough to produce an iron core with a mass exceeding 1.4 solar masses. Therefore Sirius will never become a supernova; it will end its life more quietly as a white dwarf. In addition, we can be confident that Sirius has hundreds of millions of years (at least) to go before it exhausts its hydrogen and evolves to the red-giant stage.

The most massive star anywhere near the Sun is probably Alpha Crucis, one of the bright stars in the Southern Cross. Its distance is 400 light-years, and its mass is probably about ten times that of the Sun. This makes Alpha Crucis a strong supernova candidate sometime in the future. Other stars equally or more massive are located in the Orion star-forming region at a distance of about 1700 light-years. Some of these have masses up to 25 solar masses, and they are almost sure to become supernovas. When that happens, the sight will be spectacular (like the Crab Nebula supernova of 1054 A.D.), but we expect no harm to the Earth or its population.

The problem with this survey of space near the Sun is that we are glimpsing only a moment in time. The Sun moves with respect to the stars around it, and as time passes the Sun's galactic neighborhood could change dramatically. Even more important is the time scale for the formation and evolution of very massive stars, the ones that are likely to become supernovas. These luminous stars are profligate users of fuel, compressing their entire lifetimes into a few million years. The stars that are most likely to become supernovas a few million years from now are therefore not yet born, and no amount of searching will reveal them. Precisely those objects that pose the most danger are inherently unpredictable on a time scale of several million years, similar to the time scale for large meteoritic impacts on the Earth. While statistical chances for supernovas are known, just like the chances for asteroidal impacts, specific events cannot be predicted. Our planet resides in a cosmic Las Vegas.

We should not close this chapter on a negative note, however.

Let us praise supernovas, for without them you would not be here to read this book. The atoms that make up the Earth and everything on it were created in the interiors of massive stars and ejected through supernova explosions. Without this combination of nucleosynthesis and violent disruption, the universe would be a dull place consisting of only the two primordial elements, hydrogen and helium. We are made of stardust, indeed, but more specifically of supernova dust, the heritage of generations of massive stars that existed before the birth of the Sun and planets.

CHAPTER 19

Threat from the Skies: Will a Comet Strike?

We can afford to be blasé about most of the cosmic catastrophes we have been discussing. Many of them, like the impact that created the Moon, happened long ago, when the solar system was young; similar calamities cannot happen again now that the planets have evolved. Other kinds of catastrophes affect worlds other than our own; we have nothing to fear from sulfurous, Ionian-style volcanism. Even the death of the Sun, which will certainly consume our planet in the distant future, is not a hazard to modern-day civilization. If scientists understand anything at all about stellar evolution, the Sun can no more become a red giant in the near future than a four-year-old girl can enter menopause: Both require maturation and aging. And the Sun, with its inadequate mass, is no more capable of becoming a supernova than a man is capable of becoming pregnant. Climatic changes *are* of great moment, if not for us then for our grandchildren.

But the greatest hazard of all is that civilization could be entirely destroyed any day by the unexpected impact of an asteroid or comet. We don't recommend that anyone run for cover right away. Yet, as we have seen, our planet is a target in the cosmic shooting gallery of high-speed asteroids and comets. So far, fewer than a hundred projectiles, of the presumable thousands that could terminate civilization as we know it, have been discovered in the skies.

The dinosaurs reigned on our planet for more than 100 million years before a cataclysm wiped the slate clean 65 million years ago. In comparison with the longevity of our reptilian predecessors, *Homo sapiens* has barely gotten a foothold in evolutionary history. In just the last couple of decades, however, our cultural evolution has enabled us to become aware of the nature of the threat that doomed the dinosaurs, and could doom us, as well. And we may even have the technological prowess to save ourselves from what until now could only be thought of as an act of God.

Saving our civilization from destruction by an extraterrestrial agent is not high on the list of national or international priorities today. Is this because government leaders have analyzed the situation and decided that the threat is inconsequential? Surprisingly, the answer seems to be "no." As we will recount, experts *have* considered the matter and, in fact, they concluded that the threat is serious enough to warrant awareness at the highest levels of government. However, their report was never published, so their thought-provoking conclusions apparently remain unappreciated and unheeded. The experts concluded that the probability of civilization being destroyed, in any particular year, is greater than the levels of risk our society sometimes takes very seriously; for example, they estimated that the risks of a civilization-destroying impact are several times the risk level that has been adopted by regulatory agencies as being "unacceptable," in the operation of a nuclear power plant, for a Chernobyl-type nuclear reactor disaster. That should be reason enough to look more deeply into the threat from the skies.

In June 1980, NASA's Advisory Council sponsored a symposium on "new directions" for the space agency. Some of the nation's most noted scientists, engineers, businessmen, and philosophers gathered in Woods Hole, Massachusetts, for a week-long interchange of ideas. It was hoped then, as it is still hoped today, that NASA might find a far-reaching goal for itself. The group considered various intriguing topics, ranging from solar physics to the human role in space. One subgroup considered the threat posed by an asteroidal impact. Under the rubric of "Spacewatch," such people as Luis Alvarez (who had just proposed his ideas about the Cretaceous–Tertiary extinctions), NASA head Robert Frosch, Princeton's visionary physicist Freeman Dyson, and Hewlett-Packard's Bernard Oliver wrote a short analysis of the threat. They concluded that the threat was sufficient to take action, and they proposed "Project Spacewatch" to deal with it. The idea was to build optical and radar telescopes capable of discovering the potentially threatening asteroids and comets, and then—in the unlikely event that one should be found to be on an eventual collision course with Earth—to undertake a rendezvous space mission to "give the object the proper nudge," perhaps by exploding a bomb alongside it.

NASA never selected a "new direction" for itself. But it did follow up the Spacewatch idea by asking its Jet Propulsion Laboratory (JPL) to run a workshop on the topic a year later in Snowmass, Colorado. Chaired by none other than Gene Shoemaker, the meeting brought together astronomers knowledgeable about asteroids and comets, military experts familiar with explosions, NASA engineers who knew how to plan any spacecraft missions that might be necessary for averting a disaster, and other scientists and engineers. The experts went into much more detail than the NASA thinkers had the previous summer in Woods Hole, and by September 1981 a 100-page report had been drafted.

The report was never released. There is no simple reason it was not. Some of the NASA and JPL managers were a bit leery about going public with the report. Even some of the workshop participants had doubts. For example, one scientist wanted to excise a major topic treated in the draft report: how to "nudge" any threatening asteroid away from a collision course. Since the report concluded that a big explosion would be the most efficacious approach, he worried that public discussion about putting nuclear weapons in space would be dangerously unsettling; his thinking, in the pre-glasnost years of the early 1980s, was that such discussion would be unwise and, indeed, unnecessary since the chance that any projectile might need to be "nudged" is so small. Probably the overriding reason the report was never released was that Gene Shoemaker and the other leaders of the Spacewatch Workshop simply became too busy with other projects.

The report might have created a public sensation. But its conclusions really should not surprise anyone familiar with the topics we explored earlier in this book. The draft report recounted the story of the great Tunguska event of 1908 and the then-new Alvarez hypothesis about mass extinctions. It also described such historical events as the Revelstoke meteorite, which fell in Canada in 1965 with an airburst equivalent to 20 kilotons of TNT. It asked disturbing questions about whether the spectacular phenomena associated with such fairly common smaller impacts might be mistaken by early-warning systems as a nuclear attack, and thus might accidentally trigger nuclear war. The draft report also considered how often the less common, larger asteroids and comets strike the Earth. While the average impact rates can be estimated reliably, scientists are very uncertain what the effects on civilization of such an enormous impact might be. The workshop participants attempted to address effects on the ozone

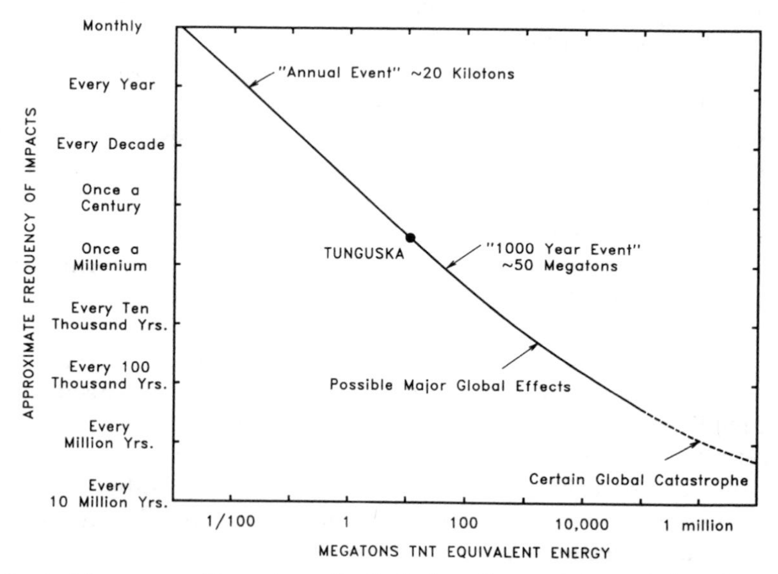

Estimated frequency of impacts on the Earth from the present population of comets and asteroids, and evidence from lunar craters. The megaton equivalents of energy are shown, as are possible and nearly certain thresholds for global catastrophe. (Clark Chapman, based on Snowmass workshop)

layer, possible tidal waves due to ocean impact, climate changes and effects on agriculture, and other topics that have been treated more thoroughly since then, in the context of nuclear winter and the Cretaceous–Tertiary extinctions.

On the one hand, such explosions as the 12-megaton-equivalent Tunguska event or the largest weapons test ever conducted above ground (59 megatons) had no major global effects. On the other hand, whatever struck the Earth 65 million years ago wiped many species from the face of the Earth. Obviously, civilization might be threatened by an insult to the environment much less catastrophic than whatever eradicated so many biological lines from the geological record. Anyone who experienced the "night the lights went out," November 9–10, 1965, learned about the fragility of our modern society; a minor failure in a Canadian generating plant near Niagara Falls triggered havoc from Boston to Philadelphia. A rather innocuous volcanic explosion in Indonesia created climate changes that caused the "year without summer" for New England in 1816, and began the exodus of farmers to the Midwest. How big a catastrophe can society survive?

The Spacewatch Workshop did *not* equate the destruction of civilization with the total extinction of the human species. Instead, it envisioned an event that might eradicate one year's agriculture and destabilize social and economic structures to the point that a new Dark Ages or even a Stone Age might result. Five, or even 90, percent of the world's population might actually survive a "civilization-destroying event." In this sense, the hazard is more like that of nuclear winter than the Cretaceous–Tertiary mass extinction. To some degree, these issues of mass psychology and estimating sociological responses to catastrophe are more nearly in the purview of science-fiction writers, like Jerry Pournelle and Larry Niven, who addressed these matters in *Lucifer's Hammer*. But the workshop participants made the best estimate they could.

Their conclusion was that each year civilization is threatened with destruction with a probability of about 1 in 300,000. That's about 1 chance in 6000 of such an impact happening during a person's lifetime (taking 50 years as a lifetime). The 1 in 300,000 chance per year is a very crude estimate; the workshop acknowledged that the chances might be as bad as 1 in 10,000 (that is 1 in 200 per lifetime) or as optimistic as only 1 in a million (1 chance in 20,000 per lifetime) or less. Finally, the draft report concluded that a dedicated telescope could detect many of the most hazardous near-Earth asteroids in a 10-year program and that larger efforts—well within technological capabilities and practical cost limits—could discover even more of them, could track their trajectories, and could "ameliorate" a potential catastrophe, if an object were discovered to be on a collision course.

We have used some very dry words and numbers to relate probabilities about the most awful imaginable catastrophe—the very destruction of civilization as we know it. Thousands of years of history could be erased in a single day, and with it the hopes for future generations, perhaps forevermore. What do these probabilities mean? Should we shrug our shoulders and watch the next TV program? After all, who ever said that living isn't a bit risky? Or should we rally the nations of the world to mount a Project Spacewatch to remove this threat from hanging over us? After all, what could possibly be better worth our concentration and effort than protecting civilization against the risks of a probably preventable accident?

*   *   *

We take risks every day. When we step out of our home and drive to the store, we run a tiny risk of being hit by a meteorite. Actually in 1954 a woman sitting in her living room was struck a glancing blow by a meteorite that crashed through the roof of her home in Alabama. Another meteorite struck and killed a dog in Egypt. But a much higher risk in our trip to the store comes from driving the automobile; although there is only a 1 in 4 million chance that we will die in a particular auto trip, all of us take so many trips that 50,000 Americans are killed each year in automobile accidents. If, instead of driving, we walk to the store, we risk getting skin cancer from exposure to the Sun. At the store, we risk buying foods with aflatoxin contamination or a toy that might choke a child. So, instead of going to the store, we might better stay at home. But 20,000 Americans die each year from accidents at home. Even drinking chlorinated water from the tap is a cancer risk, while drinking unchlorinated water poses different hazards. Should we seek to avoid all risk by remaining in bed, we risk dying of hypostatic pneumonia.

Society seems inconsistent and inequitable in its approach to reducing risks, at least in terms of the fatality numbers for various hazards. Some years ago France, for example, spent $1 million per life saved in programs designed to prevent airplane accidents, but only

$30,000 per life saved in automobile-accident prevention. Typically, regulatory agencies like OSHA and EPA believe that expenditures of many hundreds of thousands of dollars per life saved are worthwhile (similar to the costs of dialysis for a patient with kidney failure). The estimated effect of some proposed nuclear power plant rules have been estimated at $5–$10 million for each premature death averted. Meanwhile, our government spends paltry sums to protect poor migrant farm workers from some widely ignored hazards. And it has actually fostered the tobacco industry, despite the fact that cigarettes have been demonstrated to kill hundreds of thousands of Americans each year.

The apparent inconsistencies in responding to risks reflect our own individual, psychological ambivalence. We all know people who worry about food additives while a cigarette dangles from their lips, and others who drive around without seat belts but are afraid to fly in an airplane. Many of us worry about the danger of fireworks around the Fourth of July, but are complacent about using electrical appliances in our homes; yet 150 people are accidentally electrocuted for each one killed by an errant firecracker. We worry about botulism in our food and we are horrified by sensational news reports of a camper being mauled by a grizzly bear or a swimmer being attacked by a shark, but a thousand times as many people die by accidental poisoning in the home.

The science of risk assessment has grown enormously in the past two decades, in response to rising public concerns about health, the environment, nuclear power plants, airline safety, and so on. As the analysis moves beyond simple numbers to the reality of what makes people afraid, risk analysts have learned that our fears respond to some clear and sensible differences in the *quality* of the hazard, as well as to its numerical value. For example, people are far more afraid of a hazard over which they have no control than one about which they feel (however erroneously) that they do have control. This explains the widespread acceptance of risk by automobile drivers or mountain climbers compared with passengers in jumbo jets operated by invisible pilots and air traffic controllers. People are especially unaccepting of hazards they view as involuntary. Passengers on an airplane or a roller coaster don't feel in control, but at least they got aboard voluntarily. However, people subjected against their will to an environmental hazard, like contaminated drinking water, are much less accommodating. According to Chauncey Starr, one of the pioneers in risk assessment, the public may accept voluntary risks that are a thousand times greater than involuntary risks.

Another factor that affects our apprehension about a hazard is the degree to which we understand it. We feel we know the hazards involving trains and bicycles but we are afraid of radiation, unknown chemicals, and invisible microbes. Still another factor—and one relevant to this discussion—is whether the risk is catastrophic or not. We rarely think about the hundreds who die each day from the complications of smoking, bad diet, and driving while drunk, unless the victim is a friend or relative. But we all gasp in horror when an airliner crashes or when an industrial accident poisons hundreds of people living in a town. Sensationalistic news media and disaster movies feed on our fascinations and fears about horrible catastrophes.

Most dreadful of all, according to one recent study, are those hazards that raise the specter of a danger to society, to future generations. Nuclear power plants command our attention because of the potential ripple effect of a disastrous meltdown through the social and economic fabric of our society. Nuclear power is often perceived by the public as being too complex and hard to manage. Accidents are difficult to contain, threatening the uncontrolled release of invisible radioactives that could spread around the globe, inducing cancers that take years or decades to appear. Associated with radiation are risks to the unborn, to future generations. The low numerical risks aside, nuclear power plant accidents portend the breakdown of technological society. At least, that's the fear people express. And the fears are not without some justification. The risks of accident appeared to be grossly underestimated by experts before the Three Mile Island accident, and then the Chernobyl disaster happened; the experts' continuing views that the public should "calm down" about this "exaggerated" danger were once again undermined.

Governments around the world have responded with guidelines for nuclear power plants that many people feel are still too lenient, but that are actually very much stiffer than those proposed to deal with virtually any other risk. Another example is that a few publicized carcinogens, like trichloroethylene (TCE), are treated stringently by government regulations, while much more hazardous aspects of living, like bad air in our homes, go essentially unregulated. These actions all reflect the complexities of human psychology, as mediated through public discussion, private manipulation, economic power, and political action.

Where in this array of risks should we place the risk of catastrophic destruction of civilization by collision with an asteroid or comet? If the Spacewatch Workshop's best estimates are valid (and the uncertainties are great), your risk of experiencing a civilization-destroying catastrophe is:

(i) About 2000 times greater than your risk of dying from exposure to TCE at the EPA limit

(ii) About 300 times greater than your chance of dying from botulism poisoning

(iii) About 100 times greater than your chance of dying from fireworks

(iv) About 10 times the chance you will die in a tornado

(v) About a third of the risk of death by firearm accident

(vi) About 1/30th the chance you will be murdered

(vii) About 1/60th the chance you will die in an auto accident

All in all, an individual's risk of being involved in this cosmic catastrophe—whether or not he or she actually dies in it—is something like the chance a typical American will die by accidental electrocution, or several times the chance of dying in a major airplane crash. You might object and say, "But people are electrocuted every day, and nobody has been killed by an asteroid impact—the numbers must be wrong." But the numbers are roughly correct. The statistics refer to an individual's risk, and you have not been electrocuted or died in a plane crash, nor, the chances are, has anyone you know. But if an asteroid crashes, which happens *much* less frequently than an airplane crash, we might all go together. Should you worry? Rationally, you shouldn't—not if you smoke or drive or do the other things many of us do with impunity that are much more dangerous.

But the threat to civilization by an extraterrestrial impact has all the earmarks of the kind of hazard that people, and society, are hyper-concerned about. Paul Slovic, a psychology professor in Oregon, has studied attitudes toward regulation of hazards and finds fears highest for hazards that are uncontrollable, involuntary, fatal, "dreadful," catastrophic (especially globally catastrophic), and that have consequences that seem inequitable, especially if they affect future generations. Other factors augmenting fears are perceptions that a hazard is newly recognized or its risks are increasing, that it is due to unobservable agents, or that it is difficult to assess or control.

Unless we mount an effort to discover potential asteroidal or cometary impactors and start working on methods of protecting ourselves, their threat is "uncontrollable." It is certainly involuntary. Its consequences would be fatal. It is not only catastrophic but globally catastrophic. And it has the highest imaginable consequences for future generations. Also, the risk is a newly realized one and it is due to unobserved bodies (undiscovered asteroids rather than invisible microbes, in this case). This risk is literally dreadful and unimaginably horrible in its consequences. Finally, as we have already described, the magnitude and nature of the hazard to civilization are difficult to assess, and ideas about how to protect ourselves are in their infancy: Although protective techniques seem possible, they would be controversial if they were to involve nuclear weapons in space. All of these traits suggest that society might be inclined to take this threat more seriously than indicated by the modest level of numerical risk. Numerically, the probability of disaster is very low, but its consequences would be unimaginably and horribly great.

The lack of governmental response so far may be due to inadequate appreciation of the risk, because of limited publicity about it. But there is another factor we haven't mentioned yet. About a decade ago, risk-assessment experts Chauncey Starr and Chris Whipple suggested that people view some risks as "negligible" or "impossible"—hence not worth worrying about no matter how bad the consequences—if they are lower than a certain amount, perhaps a chance in a million. Maybe an annual risk of 1 in 300,000 of a catastrophic impact is somehow below our threshold of caring.

Why, then, are people worried about risks from nuclear power or food additives that are estimated by experts to be so much lower? The answer may well be that many people simply *do not believe* the experts in such cases. Certainly the perceived history of expert opinion about nuclear power plant safety has been unsettling. Conclusions of the early-1970s Rasmussen report on nuclear power planet safety were undermined by Three Mile Island and Chernobyl. General public distrust in scientists' abilities to assess the possible range of uncertainty in their results has been confirmed by studies which show that scientists and engineers are habitually overconfident in the reliability and validity of their conclusions. In addition, the public is always distrustful of studies that are perceived to be influenced by economic motivations of the involved industries. It hardly helps public confidence that some scientists employed by tobacco companies continue to deny the compelling association between smoking and some of our most serious diseases.

Whether or not the public regards a threat as "negligible" also depends on how it is perceived. Risk analysts think that some smokers ignore the hazards of smoking because, due to their addiction, they psychologically rationalize that any particular cigarette can't hurt much: Its effect is "negligible." Indeed, a smoker's increased risk from a single cigarette is less than 1 in a million. Yet smoking pack after pack, week after week, year after year adds up to 5 or 6 years shorter lifetimes for typical smokers. Failing to buckle up before a single auto trip has a similarly minuscule risk, yet one third of us will suffer at least minor injuries in a traffic accident sometime during our lives, many of which could be prevented by using seat belts. Evidently we accommodate ourselves to the most hazardous aspects of life by psychologically dividing them up into little bits and pieces. We fail to realize that "every litter bit hurts" and that the bits add up to a major problem.

Other, lesser risks command our attention because we see them as "always there." When we are swimming in warm waters, we sense that sharks are always lurking out there, somewhere. We don't want to be shredded to death, ever. Radiation seems to hang in the air perpetually; we don't mentally divide it into bite-sized pieces of hour-long exposure. How the public might respond to clear awareness of the threat from asteroids and comets perhaps depends on how it is perceived. Is the threat of impact a ceaseless menace, hanging above us in the darkness of space prepared to strike at any time? Or is an impact a singular, instantaneous, unavoidable natural catastrophe or an "act of God," as Californians seem to regard earthquakes (day in and day out, no substantial earthquakes are felt in Los Angeles, although seismologists are certain that the Big One will happen within the next couple of decades).

* * *

We've laid out the probabilities. Now we'll offer our opinion about what we think would be an appropriate response by society to this heavenly hazard. The probabilities of such a cosmic catastrophe are very small, smaller than we can really comprehend. But the stakes are truly staggering: Not only our lives, and yours, but hundreds of millions if not billions of lives, and the very future of civilization itself hang in the balance. A wise gambler sticks with the straight odds. By that measure, society should pay more attention to banning cigarettes, and stopping drunk driving, rather than expending its resources on discovering and protecting us from comets and asteroids. But the annualized threat to each individual from a cosmic impact is higher than risks from TCE, asbestos insulation, saccharin, firecrackers, or nuclear power.

The first thing we would recommend, which is cheap, is renewed thinking about the issue. The Spacewatch Workshop is nearly a decade old, its report was never released, and a lot of pertinent research has been done in the meantime. The wide range of uncertainty (risks ranging from a chance in 200 to a chance in 20,000 per lifetime) could be narrowed. The least we can do is refine our understanding of the magnitude of the threat. Provided that the best estimate of the threat to civilization remains greater than a chance per million per year, we think it would be only prudent to mount a search for the potential impactors. The cost would be a very modest fraction (perhaps 1%) of NASA's annual budget to build, equip, and operate the telescopes necessary for the search, which could be completed in less than a decade.

A similar expenditure would be more than sufficient to clarify the appropriate method and prepare for ameliorating a potential impact, in the unlikely event that one were forecast. There is no need for urgency, in our view. Certainly there is no need to ready a rocket for launch to meet the threat. At odds of 1 to 300,000 per year, we believe we can afford to wait a few years to prepare for this contingency in an orderly, intelligent way. But to let this threat slide for decades, without action, would be to treat our whole society the way a smoker treats his or her life: irrationally. If it is worthwhile to buy state lottery tickets or home fire insurance, it is sensible to invest in further knowledge about the threat from the skies. If nothing is discovered to be on a collision course, we will have lost nothing. At least we will have learned a lot more about asteroids and comets. If, instead, a doomsday impact is discovered and averted, then we will have made one of the most important contributions to human civilization in recorded history.

* * *

We human beings crave security, tranquillity, and stability in our existence. Yet we live in a hazardous world, indeed in a hazardous universe. Scientific and religious thinking of past centuries alternately portrayed our world as a uniformitarian Garden of Eden, or as a tumultuous, chaotic place where cataclysms threatened to wipe our species from existence. Most modern religions, and certainly most 20th-century sciences, have tended to move away from the catastrophist view of life. But there has been a dawning realization among scientists in many fields during the past couple of decades that the universe is less steady, less tranquil than we had imagined. Not long ago, the "steady-state" universe was a popular view of cosmology. Nowadays it is widely agreed by astrophysicists that the universe was created in a Big Bang some 15 billion years ago.

The processes set in motion by the Big Bang have important constancies and regularities, to be sure. But many stars die catastrophic deaths, to which we owe the creation and liberation of many of the chemicals of which we are made. We increasingly understand that the early history of planets may truly have involved worlds in collision. The long-accepted stability of the solar system has been challenged by the discovery of chaos. Our spacecraft explorations of other worlds have amazed us by discovering volcanism on unimagined scales, crater-saturated landscapes, and evidence for drastic climatic changes on other worlds. Studies of our own world have revealed the fragility of our ecosystem, and the potentially disastrous impact that modern civilization is having on the delicate balance of our Earth. We have become aware that evolution of biological species has occurred erratically, perhaps due to random extraterrestrial catastrophes. And, by analogy with the fate of the dinosaurs, we now perceive the very real possibility that a future impact could have similar, terminal consequences for civilization.

We do not suggest that the pendulum has swung back to the biblically inspired catastrophism of the early 19th century. Instead, we believe that the steady, objective progress of science has revealed hard evidence that unexpected, isolated, powerful events and changes are part of the natural universe. Because those cataclysms that might affect us tend to occur so infrequently compared with our lifetimes, they are difficult for us to comprehend. But when viewed from the perspective of the history of the cosmos, or the history of our planet, or even the recent epochs since human beings evolved, immense catastrophes have played a major—and sometimes dominant—role. Perhaps the vagaries of our day-to-day existence in modern society have conditioned us psychologically to accept and deal with the vagaries of the cosmos. Whether we choose to act rationally to modify our behavior and avert the potential threats remains to be seen.

Glossary

Accretion—Gravitational accumulation of mass into a planet or into a planetary precursor body, such as a *planetesimal* or a *protoplanet.*

Ad hoc—Arbitrary or without basis.

Apollo asteroid—See *Earth-approaching asteroid.*

Asteroid—Any small body (less than 1000 miles in diameter) orbiting the Sun that does not display the atmosphere or tail associated with a comet; also called "minor planet."

Astrobleme—A crater on the Earth caused by an extraterrestrial impact; a meteorite crater.

Astronomical Unit (AU)—The mean distance between the Earth and the Sun, commonly used as a measure of distances within the solar system.

Basalt—A dark-colored, fine-grained *igneous rock;* lava.

Basin—A very large impact *crater,* typically 200 miles or more in diameter, usually characterized by multiple concentric walls or rings of mountains.

Big Bang—A popular hypothesis concerning the origin of the universe, in which time, energy, and matter originated at a single place in an instantaneous "explosion."

Carbonaceous—A term characterizing a grain, rock, *asteroid, meteorite,* or planetary body in which carbon is abundant, causing the object to appear black in color.

Carbonaceous meteorite—A *primitive* meteorite made primarily of *silicates* but often including chemically bound water, free carbon, and complex organic compounds. Also called "carbonaceous chondrite."

Catastrophism—The concept that the geology of the Earth and planets has been greatly influenced by rare events of large magnitude, such as impacts by *asteroids.*

Celestial dynamics—The study of the motions of celestial objects (e.g., rotations, orbits) and the forces that affect them.

Chaos—Generally, random, disordered confusion; specifically, the new mathematical science dealing with chaotic phenomena, such as turbulence.

Chaotic orbit—A path taken by a celestial body that has been influenced by resonant gravitational forces of other bodies in such a fashion that it behaves counterintuitively by changing rapidly and unpredictably.

Chemical equilibrium—The state in which various chemicals tend not to react with each other at the ambient conditions (e.g., temperature, pressure, particular mixture of chemicals present).

Chlorofluorocarbon (CFC)—An organic compound, containing chlorine and fluorine, that leaks into the stratosphere and is involved in chemical reactions that affect *ozone*.

Comet—Any of the most primitive members of the solar system, consisting of a small body orbiting the Sun composed of ices, *silicates*, and *carbonaceous* material, which when heated generates a tenuous temporary atmosphere as its *volatiles* evaporate.

Comet nucleus—The solid part of a *comet*, typically a few miles (up to tens of miles) in diameter and probably consisting of a mixture of ices, solid *silicates*, and *carbonaceous* material.

Comet tail—The long, tenuous part of a *comet*, streaming generally away from the Sun, consisting of dust, gas, and ions.

Continental drift—A gradual motion of the continents over the surface of the Earth due to *plate tectonics*.

Cosmic abundances, cosmic composition—The proportions of chemicals characteristic of the Sun and other stars.

Cosmochemistry—Study of the chemistry of celestial bodies, especially from laboratory measurements of extraterrestrial samples, like *meteorites*.

Crater—A circular depression (from the Greek word for "cup"), generally of impact origin.

Creation science—Study of the history of the Earth and its life from the perspective of biblical stories of creation rather than from objective analysis of all available physical evidence.

Cretaceous–Tertiary boundary—The location, in a sequence of rock layers, marking the end of the older Cretaceous age and the beginning of the more recent Tertiary age, or the time (about 65 million years ago) that marks the beginning of the Tertiary.

Crust—The outer solid layer of a planet; on Earth, the upper 5 to 40 miles.

Cryptovolcanic structure—One of a group of unusual geological features, now realized to be remnants of *meteorite craters*.

Earth-approaching asteroid—An *asteroid* with an orbit that approaches the Earth's orbit to within 0.3 *Astronomical Unit;* includes the *Apollos,* which actually cross the Earth's orbit.

Ejecta—In a cratering event, the material excavated and propelled at high velocity beyond the *crater* rim.

Equilibrium condensation—An idealized model for the origin of the planets, developed about 1970, based on the assumption that planetary compositions are due to the quiescent accumulation of materials that condensed from a hot *solar nebula* as it slowly cooled under conditions of *chemical equilibrium*.

Escape velocity—The outward velocity of an object, relative to another body, necessary for it to overcome that body's gravity.

Family of asteroids—A group of *asteroids* with similar orbital elements, indicating a probable common origin in a collision sometime in the past.

Feeding zone—In the formation of a planet, the annular region from which most of the *planetesimals* came to make up the planet.

Fractionation—Preferential removal of certain chemical elements.

Fragmentation—Breaking apart by impact.

Fusion reaction—In the interiors of stars, the combining of light elements to form heavier elements, accompanied by the release of energy.

Galactic plane—The planar surface marking the orientation of the disk-shaped Milky Way Galaxy.

Geological column—The order of rock layers created during the course of the Earth's history, from older rocks at the bottom to the youngest at the top (see also *Stratigraphy*).

Geological time scale—The history of the Earth over the past 4.5 billion years, as determined from the rocks deposited in its *crust*.

Gravitational energy—The energy attained by a forming body from the infall of its constituent bodies; equivalently, the energy necessary to disaggregate a body and remove the pieces at *escape velocity*.

Greenhouse effect—The blanketing of infrared radiation near the surface of a planet by, for example, carbon dioxide in its atmosphere, producing an elevated surface temperature.

Groundburst—An explosion at the surface of the Earth, as distinguished from an underground explosion or an airburst.

Half-life—The time required for half of the radioactive atoms in a sample to disintegrate, used to measure the ages of rocks.

Highlands—The more elevated portions of a planet's surface; continents, in the case of the Earth, or the brighter, more heavily cratered terrains, in the case of the Moon.

Hypervelocity—A velocity exceeding several miles per second.

Ice age—One of the periods in the Earth's climatic history when global cooling led to the formation of extensive ice sheets over polar and even temperate land masses.

Igneous rock—Any rock produced by cooling from a molten state.

Interglacial—The period of warmer temperatures, between glacial episodes, when glaciers recede.

Iridium anomaly—An unusual concentration of the rare chemical element iridium, found at a location in the *geological column,* such as the enhancement discovered at the *Cretaceous–Tertiary boundary.*

Iron meteorite—A *meteorite* composed primarily of metallic iron and nickel and thought to represent material from the core of a melted and fractionated *parent body.*

Isotope—Any of two or more forms of the same element, the atoms of which all have the same number of protons but different numbers of *neutrons.*

Kinetic energy—The energy of an object that it has by virtue of its motion (as distinct, for example, from its thermal energy due to its temperature), which could be converted to other forms of energy by collision.

Late Heavy Bombardment—A period of cratering that occurred about 4 billion years ago, accounting for many *craters* on the Moon and the planets of the inner solar system.

Light-year—The distance it takes light to travel in one year, nearly 6000 billion miles.

Magma—Molten, liquified rock beneath a planet's surface; when it flows out onto the surface, it is called "lava."

Magnetosphere—The region around a planet in which the planet's own magnetic field, as opposed to the interplanetary field associated with the *solar wind,* dominates the behavior of charged particles.

Main-belt asteroids—*Asteroids* that occupy the main asteroid belt between Mars and Jupiter, sometimes limited specifically to the most populous parts of the belt, from 2.2 to 3.3 AU from the Sun.

Main sequence—The distribution of properties of stars of various masses (in terms of temperature and brightness) that represents the most stable, long-lived states.

Mantle—The part of a planet between its *crust* and core; on Earth, the mantle is the greatest part of the planet, with about 65% of its mass.

Mare (plural *maria*)—On the Moon, the smoother, less cratered, darker lowlands.

Mass extinction—The sudden disappearance in the fossil record of a large number of species of life, to be replaced by new species in subsequent layers. Mass extinctions are indications of catastrophic changes in the environment, such as might be produced by a large impact on the Earth.

Megaton—An explosive force equivalent to that of a million tons of TNT.

Meteorite—A rocky or metallic fragment of interplanetary debris (e.g., a piece of an *asteroid*) that survives passage through the atmosphere and strikes the ground.

Model (scientific)—A hypothetical physical explanation for a variety of observed data, often involving a number of different steps or processes.

Neutrino—A fundamental particle with no charge and little or no mass.

Neutron—A subatomic particle with no charge; together, protons and neutrons make up most of the mass of atoms.

Neutron star—A very high-density star made up mostly of *neutrons*.

Nuclear winter—A *model* for the devastating environmental consequences of a nuclear war.

Nucleosynthesis—The creation of new elements by nuclear fusion.

Oort [comet] *cloud*—The spherical region around the Sun from which most "new" *comets* come, representing objects with average distances of about 50,000 AU from the Sun, or extending about a third of the way to the nearest other stars.

Ozone—The form of molecular oxygen composed of three oxygen atoms, O_3.

Ozone hole—A continent-sized region, recently found to exist over the Antarctic during certain seasons, in the Earth's atmospheric layer of *ozone* where the ozone concentration is much reduced.

Parent body—Any larger original object which is the source of other objects, usually through breakup or ejection by impact cratering.

Periodic—Repeating regularly, with a constant interval of time between successive repetitions.

Perturbation, perturb—The gravitational effect of one object on the orbit of another.

Photosynthesis—The process whereby plants, using carbon dioxide and sunlight, produce carbohydrates and release oxygen.

Planetesimal—A hypothetical object, typically several miles in diameter, that formed in the *solar nebula* as an intermediate step between tiny grains and the larger planetary objects we see today; the *comets* and some *asteroids* may be leftover planetesimals.

Plate tectonics—The motion of segments or plates of the outer layer of the Earth, driven by slow convection in the underlying *mantle*.

Primitive—A term used to characterize an object or rock that is little changed, chemically, since its formation and hence is representative of conditions at the time of formation of the solar system.

Proto-Moon, protoplanet—A precursor body from which the Moon, or a planet, as we know it today eventually evolved.

Pseudoscience—An endeavor that pretends to be scientific, but does not use scientific methods, for example, astrology, alchemy, or fortune-telling.

Pulsar—A celestial radio source that emits periodic radio pulses, typically several times a second; a *neutron star*.

Punctuated equilibrium—A model for evolution of species involving rapid changes between much longer periods of relative stability.

Radioactive (or *radioisotopic*) *age-dating*—The technique for determining the ages of rocks or other specimens by the amount of radioactive decay of certain radioactive elements contained therein.

Red giant star—An unusually large, bright star, but with a relatively low surface temperature so that its color is reddish.

Resonate, resonance—To be "in step" with a *periodic* beat or force; specifically, a satellite or *asteroid* is said to be in resonance when its orbital (or spin) period is a simple fraction or multiple of some other planetary period that affects its motion.

Runaway greenhouse effect—A process whereby the heating of a planet leads to an increase in its atmospheric *greenhouse effect* and thus to further heating, quickly altering the composition of its atmosphere and the temperature of its surface.

Sedimentary rock—Any rock formed by the deposition and cementing of fine grains of material, usually under water.

Shocked—Affected by—or responding to—a sudden, strong force such as a *hypervelocity* impact; in particular, modified optical properties of mineral grains after passage of a shock wave.

Short-period comet—A *comet* with a period of revolution of less than 200 years.

Siderophile—A chemical element having a propensity for being incorporated into iron, or behaving like iron, so that in the process of geochemical *fractionation*, the element is often found concentrated in the same part of a planet as is iron (e.g., the core).

Silicate—A mineral containing silicon and oxygen; silicates are the most common constituents of ordinary terrestrial rock.

Simulation—An experiment in which a complex process is imitated or replicated, either in a laboratory (perhaps at a smaller scale) or in a computer, so that it may be studied more readily (see also *Model*).

Size distribution—A mathematical or graphical description of the relative numbers of objects (e.g., *asteroids*) of different sizes.

Solar nebula—The hypothetical disk-shaped cloud of gas and dust from which the planets and the Sun formed about 4.5 billion years ago.

Solar wind—The outwardly flowing stream of ions leaving the top of the Sun's atmosphere.

Spectrum—The array of colors or wavelengths obtained when light is dispersed, as through a prism; a plot of the brightness of a spectrum as it changes with wavelength—enhancements are called "emission lines," dips are called "absorptions."

Stochastic—Characterized by randomness.

Stratigraphy (or *strata*)—The study of rock layers, particularly the sequence of layers and the information this provides on the geological history of a region.

Stratosphere—The layer of the Earth's atmosphere above the *troposphere*.

Sublimate, sublime—To pass from the solid state directly to a vapor, without passing through the liquid state.

Supernova—An immense stellar explosion, which marks the final stage of evolution of a very massive star.

Tektites—Glassy objects distributed in several parts of the world, now thought to be *ejecta* from cratering events on the Earth that have reentered the Earth's atmosphere.

Terrestrial planets—Mercury, Venus, Earth, and Mars. Sometimes the Moon is meant as well, although it is a satellite, not a planet. ("Terrestrial" means Earth-like, in this case in terms of position in the inner solar system and dominantly rocky composition.)

Tidal forces—The forces within a body produced by the difference in gravitational pull, by a nearby object, on the near side of the body and that on the far side.

Tidal heating—Heating of one body by tidal friction or repeated stressing resulting from its motion within the strong tidal field of its neighbor, as in the tidal heating of Io.

Torus—A doughnut-shaped volume of space.

Troposphere—Lowest level of the Earth's atmosphere, the bottom 6–8 miles, where most weather takes place.

Uniformitarianism—The idea in geology that the landforms we see were formed through the action of the same processes that are acting today, continuing over very long spans of time.

Volatile—Boiling at a relatively low temperature; while usage of this term depends on the context, "volatile" in this book refers to substances that are liquid or gaseous at room temperature, such as water or carbon dioxide, and "very volatile" refers to those that are mobile even at temperatures far below zero, such as the ices of methane and ammonia.

White dwarf star—The collapsed stage in the evolution of a star that has exhausted most of its nuclear fuel.

Index

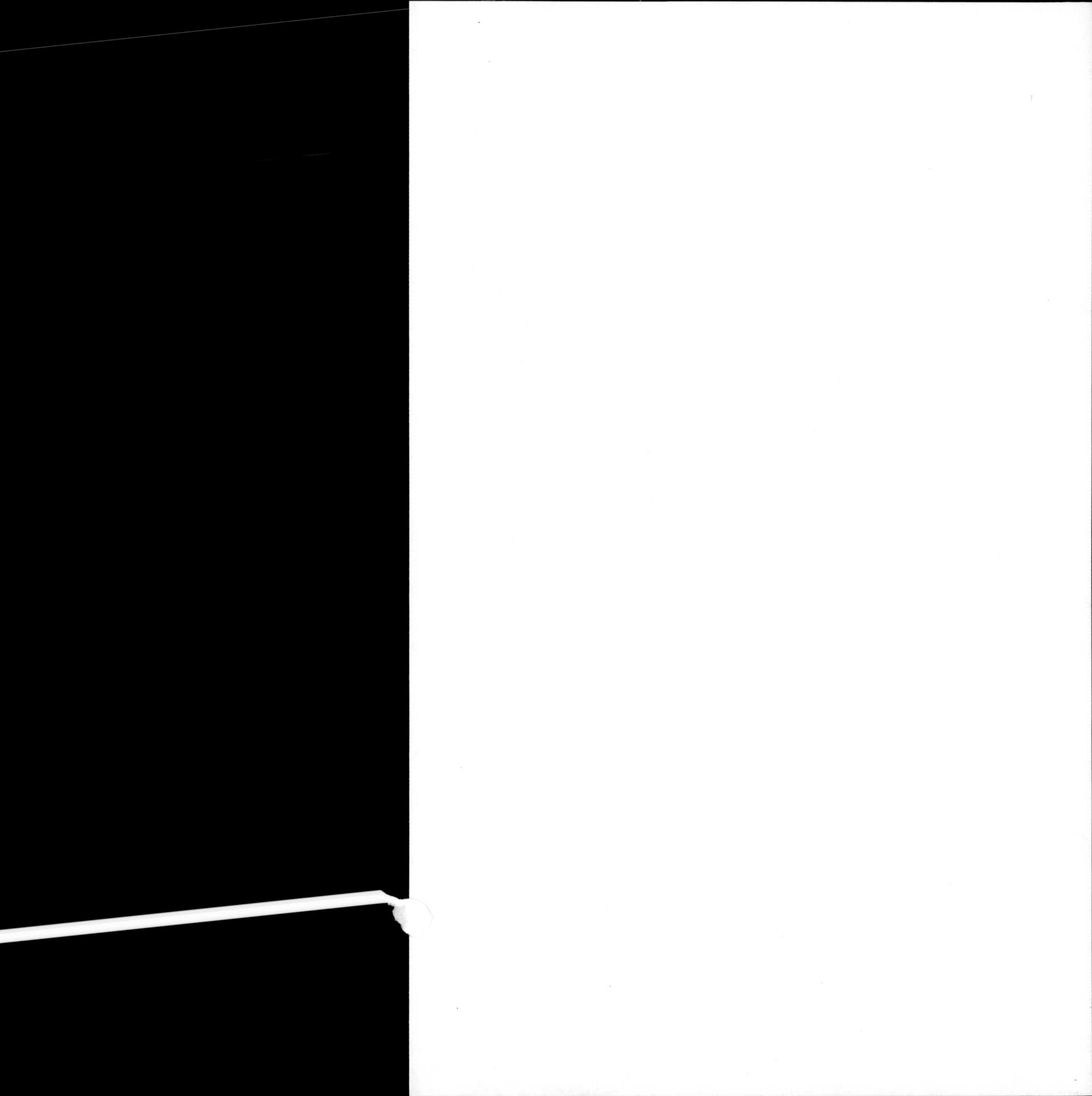